PATTERNS IN SAFETY THINKING

Dedicated to the Memory of Lorenzo Coffin
April 9th, 1823 – January 17th, 1915

Patterns In Safety Thinking

A literature guide to air transportation safety

GEOFFREY R. McINTYRE, Ph.D.

Routledge
Taylor & Francis Group

LONDON AND NEW YORK

First published 2000 by Ashgate Publishing

Published 2016 by Routledge
2 Park Square, Milton Park, Abingdon, Oxfordshire OX14 4RN
711 Third Avenue, New York, NY 10017, USA

First issued in paperback 2016

Routledge is an imprint of the Taylor & Francis Group, an informa business

British Library Cataloguing in Publication Data
McIntyre, Geoffrey R.
    Patterns in safety thinking : a literature guide to air
    transportation safety
    1.Airlines - United States - Safety measures - Bibliography
    2.Airlines - United States - Safety measures - History -
    Bibliography
    I.Title
    016.3'63124'0973

Library of Congress Catalog Card Number: 00-132578

ISBN 13: 978-1-138-25039-0 (pbk)
ISBN 13: 978-0-7546-1322-0 (hbk)

Contents

Preface

A 1992 survey by the American Society of Mechanical Engineers made the startling revelation that almost 80 percent of engineers had never taken a safety course in college nor ever attended a safety conference or a safety lecture. This naturally, has raised questions about the academic preparation of engineers who are charged with building and operating complex, high-risk socio-technical systems safely; as well as questions about the nature of the safety principles that underline their support. Traditionally, "safety" has been treated as part of the overall design process rather than being the focus of a specific discipline of study. At the end of this century and the next one, these questions are at the heart of American civilization's thinking and communicating about "safety."

Patterns In Safety Thinking: A Literature Guide To Air Transportation Safety, was written in response to an expressed need by aviation safety management trainees for a practical and *concise* educational supplement to the safety literature as it relates to the theory and practice of U.S. transportation safety. This essay will examine and classify the major books and select journals on transportation safety and where appropriate, provide real world, practical illustrations of the theoretical concerns. It is primarily designed to educate aviation transportation safety practitioners on the broad outlines of the history, theory and practice of "safety management" in the United States. Several aviation safety trainees under the time-sensitive pressure of typical training programs have expressed the need for *concise* reference material that would provide a broad orientation and conceptual overview of safety management. The essay meets that requirement. As a potential educational supplement for undergraduate engineering students, the essay serves as a guide to stimulate inter-disciplinary thinking about "safety." It should also be of assistance in meeting public policy objectives to better communicate aviation safety risk to the public. The essay takes readers on an historical journey that explains how the streams of several academic disciplines, as well as the safety actions of individuals, have merged to create the omni-disciplinary nature of "safety."

This essay classifies the transportation safety literature into a theoretical framework of distinct and apparently self-reinforcing "schools" of safety management thought: The "Tort Law School," the "Reliability Engineering School," and the "System Safety Engineering School." In addition to this categorization of the literature, it has provided a context with the inclusion of new and practical, real-world illustrations drawn largely from the aviation sector. The goal of the essay is to educate rather than to train safety engineers, students and practitioners. In general, both students and practitioners are faced with a bewildering array of highly technical, specialized and esoteric literature drawn from several disciplines, including the cognitive and mathematical sciences, probability theory, risk management, engineering, organizational theory, tort law, and the political economy and philosophy of regulation. This omni-disciplinary character has impacted the practice of safety as an integral component of corporate management and governmental regulatory actions.

Patterns In Safety Thinking, was conceived as an educational supplement to safety management training courses for aviation safety practitioners. Generally, aviation safety training courses are very intense, time sensitive, cookbooky and largely concentrate on the application of standard reliability engineering tools and techniques of modern probability theory. Many System Safety theoreticians and practitioners believe that the challenges posed by software-intensive systems will require a broadening of the scope of aviation safety training to include developing an ability to asking the right questions when *thinking* about safety. In 1978, there emerged from the safety literature new thinking by sociologists, organizational theorists and psychologists that furthered managerial understanding of the System Safety discipline. This work has examined some evolving transportation safety education, risk management and communication challenges of the 21st century.

The century-old theme that safety is *everyone's* concern began in 1874 with the actions of Lorenzo Coffin, the almost forgotten railroad safety advocate and champion of the *Railroad Safety Appliance Act, 1893.* Coffin would later serve as the role model for Ralph Nader, the 20th century's most famous safety advocate. The essay, where relevant, will also reference significant technological advances and on-the-job aviation practices of this "practical discipline." It also examines the individual actions of "ordinary" human beings that have impacted, and continue to impact safety. The tragic consequences of accidents and unsafe conditions on our lives suggest that safety is indeed *everyone's* concern. Thus, at the beginning of the 21st century, "safety," far from being viewed as the sole domain of design engineers and risk managers with their market-based models of economic rationality, has become the entrance upon a quite incalculably wider stage that has coalesced into unique patterns of thinking about "safety." This essay should also be of some assistance in meeting U.S. public policy objectives to further educate consumers about the safety of the air transportation system. It provides a framework for analysis and discussion of the 1990s issue of communicating aviation safety risk to the public. The airspace system *daily* handles more than 174,000 takeoffs and landings at airports across the nation, and *routinely* carries approximately 1.7 million passengers *safely* to their destinations. The White House Commission on Aviation Safety and Security has assigned a top priority to reducing aviation's accident rate.

Like the Roman God, Janus, "safety" may be portrayed as facing two ways: backward to the past and forward to the future. In transportation, after an accident occurs, the National Transportation Safety Board conducts an investigation to determine the cause (s) and what must be done to prevent a recurrence. In the aviation sector, this approach has been known as "Fly-Fix-Fly." Thus, the dreadful trade of accident investigation looks backwards and hopes to eliminate future sources of trouble. In contrast, the "System Safety" approach could be described as "forward looking" to the future with a focus on *accident prevention*. This involves a before-the-fact process that may be characterized as the identify-analyze-control method of safety to enable us to answer the question: How do we *know* that a system is safe? A central thesis of this essay is that safety is more than the absence of accidents. Safety is defined as the goal of transforming the levels of risk that inheres in all human activity. Such a view of "safety" challenges the intellectually entrenched tendency to communicate future risk likelihood based *solely* on studies of past accidents.

Acknowledgements

I have been fortunate to have received assistance, advice and encouragement in finishing this essay. I am especially grateful to my Federal Aviation Administration (FAA) colleagues: Steven D. Smith for his assistance, encouragement and valuable review of an early draft as well as technical assistance in manuscript formatting. My special thanks also to Dave Balderston for his comments on an early draft; and, to Michael Allocco, former Vice President of the System Safety Society for encouraging me to further develop my idea to classify the safety literature into a framework that is relevant to its teaching. My thanks to Jack Wojciech, who worked as a Reliability Engineer and introduced me to the discipline. Thanks also to Daniel C. Hedges and Jerold Feinstein for their encouragement. Special thanks also to David Gleave of Aviation Hazard Analysis, for his encouragement.

I also wish to thank the following for their valuable suggestions and constructive criticisms and suggestions for improving a prepublication draft: Dr. Najmedin Meshkati, Director Continuing Education, University of Southern California; Dr. James Luxhoj, Department of Industrial Engineering, Rutgers University; Dr. Nick Pidgeon, University of East Anglia, United Kingdom (Professor Pidgeon was formerly of the Department of Psychology, University of Wales); Mr. Terry Kelly, Manager, Safety Policy, NAV CANADA, Canada; and Dr. George Chang, retired Professor of Aeronautics, Embry-Riddle Aeronautical University.

My thanks to Amy Tursky and Frank Reilly, Librarians at the U.S. Department of Transportation.

1   Introduction

This essay will examine and classify the major air transportation safety related books and select journals on transportation safety and where appropriate, provide real world, practical illustrations of the theoretical concerns. The essay is primarily designed to educate aviation transportation safety practitioners and aviation university students on the broad outlines of the history, theory and practice of "safety management" in the United States. Several aviation safety trainees under the time-sensitive pressure of typical training programs have expressed the need for *concise* reference material that would provide a broad orientation and conceptual overview of the safety management discipline without having to resort to an entire shelf of books.[1] As such, *Patterns In Safety Thinking: A Literature Guide To Air Transportation Safety,* meets that requirement. Many of the textbooks and reports that were reviewed for this essay ranged in size from 260 pages to in excess of 600 pages. Within the confines of this survey, a full and comprehensive literature review to include scholarly journals is neither practical nor desirable. In part this would have resulted in an overly ambitious project that could easily exceed 1,500 pages--defeating the original purpose and intent of this essay, as well as the fact that not all of the risk management issues addressed by researchers are germane to aviation safety. The safety literature has drawn from among many disciplines including the behavioral and natural sciences and these incorporations are continuing to enrich its intellectual evolution and how we think about "safety management." As an educational supplement for engineering students, the essay may serve as a guide to stimulate inter-disciplinary thinking about safety. The book may also partially meet public policy objectives to further educate air travelers about safety and further aviation safety risk communication. Transportation has an enormous impact on the U.S. economy. According to one estimate, in 1995, transportation-related goods and services accounted for approximately 11 percent of United States Gross Domestic Product.[2]

The safety management literature has evolved from several incorporations from the behavioral sciences, engineering and mathematical sciences/quantitative methods. These disciplines have clearly defined various "emphases" or "schools" within the boundaries of the safety discipline and have led the author to create a taxonomy of key literary contributions. Such a classification of the books on safety that is relevant to its teaching has never been undertaken.[3] Research into the related books on safety indicated that there are good reasons why a study such as this had never been undertaken. For certain, regardless of how one were to qualify or hedge an approach such as this, it is axiomatic that, this is at best a perilous undertaking since there appears to be no one best way to view the literature. Many persons populate the safety literature and naturally, in describing the citations from the literature it does not necessarily mean that controversial views are shared. Rather, its purpose is to educate and provide concise reference material on the well-documented historical progression of the major scholarly safety literature as it relates to the theory and practice of U.S. transportation safety management.

"Safety" as a discipline has evolved from several divergent and conflicting paths and disciplines that were the result of polar patterns of thought and actions. From this omni-disciplinary character, this essay identifies various intellectual emphases or "schools of thought" - a "Tort Law School," the "Reliability Engineering School," and the "System Safety Engineering School." There is also extensive literature concerning Occupational Safety Law, Insurance, Risk Management, Product Liability, Environmental Law and Safety, and Security. However, to keep within the purpose and scope, those subject areas are only referenced insofar as they impact the aviation transportation safety emphasis. It appears that a concise summary of the historical origins and development of this "practical discipline" should help students and practitioners understand the philosophical context and the often-confusing viewpoints and the uses of terminology. Aviation safety practitioners seeking to gain a wider appreciation of the System Safety Engineering discipline and how it has evolved from "Reliability Engineering" should find this textbook useful.

A Century of Safety Concerns

This essay spans over a century of U.S. societal thought, actions and concerns over transportation safety. We begin our study with the activities of Lorenzo Coffin, the 19th century railroad safety advocate and champion of the *Railroad Safety Appliance Act, 1893*. As early as 1874, his was the pioneering voice for the merging of the two streams of safety technology and government policy control. Chapter Two also includes the activities of

Ralph Nader, the 20th century's celebrated crusader for automobile, highway and, increasingly, air safety and the "Tort Law School" of transportation safety and government regulation. Chapter Three progress to the newly emerging post Second World War "Reliability Engineering" profession. We also examine the rise of Probabilistic Risk Assessment and Human Reliability Assessment with its emphasis that human behavior can be mechanically modeled, analyzed and understood. The apparently inwardly regenerative nature of safety further gave rise in Chapter Four to the emergence in 1962 of the "System Safety Engineering School." The concerns of this school are over our ability to safely design software-intensive, electro-mechanical systems that characterize the modern air and space transportation era, and the overall safety of complex, socio-technical systems.

We begin with our classification of the literature of the "Reliability Engineering School" with publication in 1968 of Martin Shooman of Brooklyn Polytechnic Institute, landmark text, *Probabilistic Reliability: An Engineering Approach.* The book was designed for college and industrial courses in what was then the newly emerging field of Reliability Engineering. However, it should be emphasized that the roots of "Reliability" go further back to the early (1911) work of Frederick W. Taylor's: *The Principles of Scientific Management.* He is often called the "Father of Scientific Management." Taylor's dominant concern was efficiency and the best and cheapest way of accomplishing routine work. Also the experiences of America's early industrial assembly line production methods are rooted in reliability. For Shooman, the need for reliable equipment and reliability analysis became apparent at the close of the Second World War. The problems of maintenance, repair, and field failures became severe for the military equipment used during the War. In the late 1940s and early 1950s, Reliability Engineering became the newest engineering field. This new field came about primarily due to the complexity, sophistication, and automation inherent in modern technology. The fields of communication and transportation were the first to witness rapid growth in complexity as equipment manufacturers adapted advances in electronics and control systems. The field came to a focus and began to develop when people agreed that the proper definition of reliability was in terms of the probability of success. This decision marshaled the powerful techniques of modern probability theory behind the growing field of reliability.[4] It is beyond the scope of this text for a treatment of the complex subject of the laws of probability. However, Shooman's early text does contain explanations on its analytical tools as well as an excellent bibliography.

One consequence of this new pattern in safety thinking was the further evolutionary spiral to a new branch of Probabilistic Risk Assessment (PRA) within the "Reliability Engineering School." By 1975, in a continuing evolutionary spiral, the drive to an understanding of the likelihood of hardware malfunctions and errors led to the adoption of PRA (pioneered by the commercial aviation industry) by the nuclear and other high-risk industries. Then there was a natural migration to focus attention on Human Reliability Analysis. Suffice it to say that for the "Reliability Engineering School," the circle appeared completed with the migration from mechanical reliability to perfected methods and techniques for predicting human reliability.

According to Roland and Moriarty,[5] the modern discipline of System Safety evolved in 1962, with the dawn of the space transportation era. The growth and development of the System Safety approach to *accident prevention* was created by the publication of safety standards, specifications, and requirements, as well as operating instructions. The System Safety concept calls for safety analyses and hazard control actions beginning with the conceptual phase of a system and continuing through the design, production, testing, use, and disposal phases, until the activity is retired.

A Golden Decade

The period, 1965-1975, was truly a Golden Decade for safety. During that period, there was a profound experience in the demand for "safety" and the laying of a broader legal and theoretical framework for transportation safety in particular. The emergence of Ralph Nader as the undisputed leader of the public safety advocacy movement with publication in 1965 of his book, *Unsafe At Any Speed,* focused national attention on the issue of automobile safety defects, regulatory inaction and the potential for further increasing the carnage on the Nation's highways.[6] An updated version of *Unsafe At Any Speed,* was published in 1972. For Nader, the experience of the mounting four decade-statistics on automobile deaths and the lack of accountability and publicly available information or regulatory requirements for a safe, non-polluting automobile, meant that it was time to "bring the industry to justice."

During the Golden Decade, air transportation safety also came under scrutiny. However, it should be emphasized that air safety concerns go back further to the *Air Commerce Act of 1926* that in effect introduced the regulatory role of the Federal government. During the 1930s, the emphasis on industry-government cooperation on the standards for civil aviation intensified. For example, in 1935, in a remarkable development of industry-government cooperation, the Radio Technical Commission for Aeronautics (RTCA) was organized as a forum where government and aviation industry representatives gathered to develop consensus performance standards for "black box" airborne equipment. Armed with those "Minimum Operation Performance Standards," manufacturers were now able to build a product with the assurance of obtaining regulatory certification. The June 30th 1956 mid-air crash over the Grand Canyon was the first commercial aviation disaster to claim more than 100 lives. It exposed the safety inadequacies of the air traffic control system and ultimately led to the *Federal Aviation Administration (FAA) Act, 1958.*

In 1967, the book, *Airline Safety Is A Myth,* was published by Captain Vernon W. Lowell, a command pilot with Trans World Airlines for over 22 of his 28 years of airline piloting. Captain Lowell, in cooperation with the Airline Pilots Association, provided readers with actual case histories that in effect called for the need for preventive action by the airlines, aircraft manufacturers, the federal agencies with responsibility in this area and the congress. One case study in Captain Lowell's book is of the Boeing 727, which had been involved, in four fatal accidents within six months. The book documents the misguided role of the then investigative Civil Aeronautics Board in placing blame on "Pilot Error" for those accidents, notwithstanding the fact that there had been known problems with asymmetric reverse thrust on that type aircraft.[7]

John Godson's 1970 book, *Unsafe at Any Height,* called attention to the increasing number of air crashes and a public perception that air crashes were "unavoidable." The author cited the "culpably neglected" standards of air safety, ineffective accident inquiry boards that did nothing to avoid future accidents and "how tragically slow the aviation industry has been to learn from experience." The book contained ominous warnings about future accidents with the approaching era of jumbo jets. Godson acknowledged that a totally accident-free aviation world is not expected. However, his concern was "about those many areas of civil air transport where neglect and inadequacy are glossed over and condoned by inaction year after year... we are dealing not with merchandise but with human lives; and a company which takes part in this kind of business must expect to have to observe the most rigorous standards."[8]

The role of the electronic and print media further served to raise public consciousness about the terrible toll of accidents and the need for more attention to safety, produced a virtual tsunami of safety legislation. This truly Golden Decade for safety concerns at all levels of American civilization saw the emergence of new patterns in safety thinking, worthy of a great industrial and information-driven society, as reflected in congressional passage of landmark legislation. A partial list includes: the National Traffic and Motor Vehicle Safety Act, the Highway Safety Act, and the Department of Transportation Act, the Coal Mine Health and Safety Act and the Environmental Protection Act, the Boat Safety Act, the Noise Control Act, the Consumer Product Safety Act, and the Occupational Safety and Health Act. The rush to legislate safety following catastrophes has been a recurring pattern of congressional safety actions documented by Brenda McCall in her book, *Safety First at Last*.[9] Passage of the above-cited legislation ranging from the environment to occupational health and safety, consumer product safety, transportation safety etc would serve as a catalyst for the transformation of society's ideas about "safety" with far-reaching implications for the pivotal role of communicating safety risk information to the public in the conduct of domestic and international trade. The theme of safety risk communication is one that two decades later would dominate the issue of aviation safety.

During this Golden Decade, the National Safety Council conducted two symposia on safety issues: The first symposia (1966) provided an overview of occupational safety measurement practice, the second (1970) concentrated on various aspects of measurement and safety performance.[10] A number of influential textbooks were written and used in expanding University-level Professional programs in "safety." By 1969, the quest for a "science" of accident prevention, led the influential safety theoretician and practitioner, H.W.Heinrich, *Industrial Accident Prevention,* to formulate certain basic principles or "Axioms of Industrial Safety," that would be applicable to a safety management approach. Essentially, these axioms dealt with the theory of accident causation, the man/machine interface, with management's role, and the relationship between the economics and efficiency of safety.[11]

Heinrich's textbook provides an excellent summary of the background of industrial safety. In it, he has referenced the founding of a number of safety institutes including, Northwestern University Traffic Institute in 1936. He has also documented the role of insurance liability in protecting employers against legal action brought by injured workers. Heinrich viewed the government's intervention as a key role in creating the social mechanisms, by which safety technology is implemented by fixing accountability, allocating costs so that it becomes desirable or profitable for someone to devote resources to preventing loss. For him, the nature of this intervention was not to command insurance companies to prevent accidents in the workplace. Heinrich believed that it was worker's compensation that made it profitable for insurance companies to get involved in loss prevention. Without such a mechanism, the impetus for safety is nonexistent.[12] Safety became "everyone's business."

At around this time, another influential safety theoretician, Dr. John V. Grimaldi, in his text, *Safety Management*, developed the principles and practice of safety management that led to widespread adoption of the term "Safety Manager." This gradually supplanted the earlier title "Safety Engineer" which had proven to be a misnomer, since many did little or no true engineering work. Their primary task was to administer injury reduction programs. The text also has documented the evolving nature of governmental regulatory Worker Compensation laws and their impact in providing motivation for U.S. business to implement safety programs as an integral part of "accident" prevention programs rather than waiting for government intervention. This emphasis on reducing work injuries led naturally to standardization of methods of compiling work injury data. In 1937, the American Standards Association (now the American National Standards Institute (ANSI)) published material that defined worker injury in accordance with Worker Compensation Laws. The U.S. Department of Labor was created in 1913. The Bureau of Labor Statistics (a direct descendant of the Bureau of Labor, 1884) had the task to gather and disseminate labor data including the nation's work injury experience. In 1934, the Bureau of Labor Standards was established to develop desirable standards for industrial practice. New York University's Center for Safety Education was founded in 1938 by an annual gift from the Association of Casualty and Surety Executives.[13]

Toward the end of the Golden Decade, Willie Hammer, a Professional Engineer, emerged as the leading authoritative voice on safety management engineering. Hammer's 1972 text, *Handbook of System and Product Safety,* [14] laid the intellectual foundation for the movement away from a failure-centric perspective to a system-wide emphasis. That work held enormous implications for our understanding of accidents and safety risk management. No discussion of the System Safety discipline can be conducted without reference to Willie Hammer's pioneering textbook. The emphasis on industrial labor relations and safety led to the 1976 publication of another of Willie Hammer's classic texts, *Occupational Safety Management And Engineering.* Hammer observed that cost and regard for human life and well being motivated United States Steel Company in 1908 to begin "the first formal corporate safety program." He defined safety as "freedom from hazards" and postulated The *Law of Safety Progress*: An unsafe product will bring on corrective action or drive its producer out of business, thereby raising the safety level of all such products. However, Hammer conceded that it is practically impossible to completely eliminate all hazards. Safety is therefore a matter of relative protection from exposure to hazards: the antonym of danger. Hammer's work served as an inspiration for many authors.

In 1974 the book, *The Titanic Effect* by Kenneth E.F. Watt, an ecologist writing about economic problems coined the phrase "*Titanic* effect," to describe the human tendency to ignore public warnings as "unthinkable" when they were outside the range of past experience. Consequently, the appropriate countermeasures were not taken. Watt's examination of history, concluded with an important generalization: [15]

> The magnitude of disasters decreases to the extent that people believe that they are possible, and plan to prevent them, or to minimize their effects.

In general, it is worth taking action in advance to deal with disasters because the costs of doing so are generally lower when measured against the losses if no such preventative action took place. Over the years, safety specialists have agreed on the cliché: "If you think safety is expensive, try having an accident," because accidents are involuntary and unscheduled expenditures. Ironically, the phrase "*Titanic* effect" and the evidence that major accidents are often preceded by a belief that they cannot happen, would be validated fifteen years later with the 1989 *Exxon-*

Valdez oil spill in Prince William Sound, Alaska. Consider the reaction of two U.S. Coast Guard Admirals: Vice-Admiral Robbins, when he first heard of the accident said, "That's impossible, we have the perfect system," and Rear Admiral Sipes, Chief of the Coast Guard's Office of Marine Safety: "to us it was inconceivable."[16]

Transforming Ideas about Safety

In the United Kingdom, the 1978 publication, *Man-Made Disasters*, by the sociologist Barry Turner, was the seminal work on disaster analysis.[17] It is regarded as the very first comprehensive theoretical analysis of organizational vulnerability to technological disasters. At its publication the model was "some five to ten years in advance of, and provided much of the foundation for (later) work that has contributed to our present understanding of industrial catastrophe and crisis as managerial and administrative in origin."[18] Based upon a systematic qualitative analysis of British accident inquiry reports over a 10-year period, Turner observed that "disasters in large-scale technological systems are neither the product of chance events, nor Acts of God. Nor can they be described purely in technological terms. Rather, disasters 'incubate' as a series of unintended human and organizational factors (such things as slips and mistakes, fallible management decisions, organizationally amplified errors, and inappropriate assumptions about safety and danger) build up, often over a period of many years."[19] Turner defined a *disaster* not by its physical impacts but in sociological terms, as:

> A significant disruption or collapse of the existing cultural beliefs and norms about hazards, and for dealing with them and their impacts.

Turner observed that all organizations possess a variety of cultural beliefs and norms regarding the management of hazards. These are generally either laid down in rules and procedures, or culturally taken for granted work practices. "Over the course of the disaster incubation period, the latent errors and events that build up are at odds with the prevailing assumptions about hazards, making the system vulnerable to catastrophe."[20] This text raised for the *first time* many key issues long since taken as axioms in the emerging theoretical understanding of organizationally induced crises and vulnerability to accidents. A second edition, published in 1997, with Professor Nick Pidgeon, School of Psychology, University of Wales, has been updated to include topics of organizational learning (how blame gets into the system and prevents learning to occur) and culture that was in progress at the time of Barry Turner's death.[21] These issues are further examined under the human, organizational and environmental factors of system safety in Chapter 4. Professor Pidgeon has noted that Turner's original model has been built on by numerous others including Perrow, 1984; Shrivastava, 1987; Reason, 1990; Pate-Cornell, 1993.

Professor Diane Vaughan of Boston College wrote a foreword to the second edition (1997) of *Man-Made Disasters* and expressed her marvel for the late Turner's prescience and accomplishment in setting his core ideas:

> I had become aware in the 1970s that there was a potential store of information about administrative failures and shortcomings in reports of public inquiries into large-scale accidents and disasters and I concentrated this perception into my doctoral thesis, *The Failure of Foresight,* which was written at Exeter, in the UK, and formed the basis of a book, *Man-Made Disasters.*

She noted that the simple, imaginative, alliterative phrase, *The Failure of Foresight,* captured the essence of Turner's analysis which formed the basis of his 1978 book. "Man-made disasters not only *had* preconditions, but those preconditions had characteristics in common: long incubation periods studded with early warning signs that were ignored or misinterpreted. Man-made disasters were distinguished, not only by the institutional, organizational, and administrative structures associated with them, but by their *process.* Disasters were not sudden cataclysmic events; they had long gestation periods." Professor Vaughan also noted that *Man-Made Disasters* had not attracted the attention of mainstream American sociology, and neither Charles Perrow nor James F. Short, a former President of the American Sociological Association, cited Turner's research.

In the United States, in the wake of the frightening nuclear power accident at Three Mile Island on March 28,1979, 10 miles southeast of Harrisburg, Pennsylvania, there emerged a new thinking about safety. The Yale University organizational theorist, Charles Perrow, in his book, *Normal Accidents* (1984) made the point that high-risk systems have some special characteristics (interactive complexity and coupling) that makes

accidents in them inevitable, even "normal."   Perrow's book was a
watershed in new thinking about safety because, in addition to the usual
causes of accidents—some component failure, which could be prevented in
the future—was added "a new cause: interactive complexity in the presence
of tight coupling, producing a system accident. We have produced designs
so complicated that we cannot anticipate all the possible interactions of the
inevitable failures."[22] For Perrow, despite the fact most high-risk systems
have some special characteristics, beyond their toxic or explosive or genetic
dangers that make accidents inevitable, it is still possible to analyze these
special characteristics and thus gain a much better understanding of why
accidents will always occur in these systems.   That knowledge enables
policy makers to better argue that certain technologies should be
abandoned, and others, which we cannot abandon because we have built
much of our society around them, should be modified. "Risk will never be
eliminated from high-risk systems…however, we might stop blaming the
wrong people and the wrong factors, and stop trying to fix the systems in
ways that only make them riskier."[23]   With regard to the safety of air
transportation, Perrow cites the fact that the extensive operating experience
gained in several decades of flying, prevents the aircraft system from being
more risky.   However, "for the commercial success of air transport,
accidents must be reduced.   But the hard core of system accidents, while
small, will probably not get smaller.   This is because with each new
advance in equipment or training, the pressures are to push the system to its
limits."[24]   Further tragic accidents or "system accidents" in 1986 (The
Challenger Space Craft and Chernobyl to name two) and Turner's
pioneering work, followed by Perrow, have inspired leading cognitive
scientists, such as James Reason, at the University of Manchester, England,
to make the study of errors a legitimate pursuit. In *Human Error*,[25] Reason
.concluded that managerial weaknesses in complex technological systems
facilitated predictable and unintentional human errors to penetrate defenses
resulting in catastrophic loss. The 1997 publication of another of Reason's
books, *Managing The Risks of Organizational Accidents*, [26] provided safety
practitioners with the tools to predict where the breaches in risk
management defenses might exist.

       In a world of news media stories about structurally sound and
"certificated" aircraft that plummet to earth; automobiles suddenly
accelerate; supertankers run on the shoals in calm weather; and medical
machines designed to save lives, have actually maimed and killed patients,
how do we describe our safety knowledge and practices?   What are the

theoretical and practical foundations of managing risks associated with all human activity? These bizarre accidents and incidents have led aircraft designers and manufacturers to require that design engineers, as well as safety managers should be equipped with additional material *to help in asking the right questions* about safety risk management.[*]

The thesis of Steven M. Casey's book, *Set Phasers On Stun: And Other True Tales of Design, Technology, and Human Error*, is that many "human error" accidents should be more aptly named design-induced errors. Design-induced errors reflect incompatibilities between the way things are designed and the way people actually perceive, think, and act. From an air transportation safety perspective, the computer related accidents of the Airbus A320, June 26, 1988, at Mulhouse Habsheim, in Alsace, France, and the subsequent, February 14, 1990 Indian Airlines accident, at Bangalore Airport, for the first time, drew public attention to the risks of automation. The issue of automation and the role of the human operator to adapt to unforeseen circumstances would anchor the debate on air safety. Dr. Casey's book has been adopted as a supplemental reading text by many colleges. The text is also being used as supplemental reading in a wide area of courses, including human factors and accident analysis, computer science, system safety and psychology. The author has presented twenty factual accidents "to serve as vehicles for thinking about and understanding what happens when designers of products, systems, and services fail to account for the characteristics and capabilities of people and the vagaries of human behavior."[27]

A decade after Perrow's influential work, a political scientist at Stanford University, Scott D. Sagan, published in 1993, *The Limits of Safety: Organizations, Accidents and Nuclear Weapons.*   Sagan's work challenged the beliefs of scholars who have held that key factors that lead to high organizational reliability—redundant back-up systems, personnel discipline, and trial-and-error learning—have not produced a safe nuclear arsenal. For Sagan, John von Neumann addressed a major safety challenge when he asked: "Is it possible to build "reliable systems from unreliable parts?"[28] This is a fundamental challenge to builders of socio-technical systems steeped in the thinking patterns of the "Reliability Engineering

---

[*] Discussions with Mr. David Allen, Chief Engineer, CNS, (Communication Navigation Surveillance) – Air Traffic Management, Boeing Commercial Airplane Group. He has required newly employed engineers in his division to read Steven M. Casey's book: *Set Phasers On Stun: And Other True Tales of Design, Technology and Human Error*. (Santa Barbara, CA: Agean Publishing Co., 1993).

School," and its reliance on redundancy. The topic is examined at length in Chapter 3. A fundamental premise of "Reliability" is that all bad things result from failure; equipment reliability equates with safety; and safety is reinforced by redundancy. What has emerged from societal experiences with accidents in computer-controlled systems, are *new patterns in safety thinking*: electromechanical systems do not necessarily have to "fail" to produce a catastrophic event.

In 1995, Nancy Leveson, Professor, Department of Aeronautics and Astronautics, Massachusetts Institute of Technology, published *Safeware: System Safety and Computers.* This 680-page textbook contains no footnotes but lists a bibliography of 358 references. It is a major contribution to our understanding of the role of software in high-risk systems and the importance of integrating software safety efforts with system safety engineering. Her basic focus is that safety is not a property of software, but a property of the system. The book is a mix of system analysis, system safety, software techniques, management, and organizational approaches. The book echoes many of the themes that Willie Hammer and other early System Safety professionals have written about. It examines what is currently known about building safe electromechanical systems and looks at the accidents of the past to see what lessons can be applied to new computer-controlled systems. Accidents are not the results of "unknown scientific principles but rather of a failure to apply well-known, standard engineering practices. Accidents will be prevented by not only technological fixes, but also will require control of all aspects of the development and operation of the system."[29] In 1999, the publication of *System Safety: HAZOP and Software HAZOP,* provided safety practitioners with a detailed discussion and practical examples of the hazard and operability (HAZOP) technique to the identification and analysis of hazards in software-based systems.[30]

Leveson's book also provided an insightful reference to the problem of using scientific methods and arguments to answer what are fundamentally *not* scientific questions. She quotes, Alvin Weinberg, former head of the Oak Ridge National Laboratory in an attempt to separate out and clear the moral, ethical, philosophical, and socio-political questions that cannot be answered by algebraic equations or probabilistic valuations:[31]

> Many of the issues that lie between science and politics involve questions that can be stated in scientific terms but that are in principle beyond the proficiency of science to answer. ... I propose the term "trans-scientific" for such questions. ... Though

they are, epistemologically speaking, questions of fact and can be stated in the language of science, they are unanswerable by science; they transcend science....In the current attempts to weigh the benefits of technology against its risks, the protagonists often ask for the impossible: scientific answers to questions that are trans-scientific.

Clearly, the term "trans-scientific" will prove a useful concept in understanding the limitations of algebraic equations or probabilistic valuations and a civilization's quest for considering the social costs of environmental degradation and the safety of socio-technical systems. It is a theme that we shall examine in greater depth when we discuss the "Reliability Engineering School" and issues in risk assessment. Leveson's book also contains appendices that are rich in examples of designed-induced errors, including, Medical Devices: The Therac-25 Story. Between June 1985 and January 1987, this computer-controlled radiation therapy machine overdosed six people. These accidents have been described as the worst in the 35-year history of medical accelerators. This has also been cited throughout the literature as the best example of the fact that fatal accidents can and have been caused by a unit functioning as designed without failing to meet its design requirements. A lesson to be learned is that: [32]

Focusing on a particular software 'bug' is not the way to make a safe system. Virtually all complex software can be made to behave in an unexpected fashion under some conditions. The basic mistake here involved poor software engineering practices and building a machine that relies on the software for safe operation.... The particular coding error is not as important as the general unsafe design of the software.

Professor Leveson's examination of safety in civil aviation and the case of the DC-10 Cargo Door was drawn from the 1976 investigative report, *Destination Disaster: From the Tri-Motor to the DC-10—The Risk of Flying*. That material is examined in fuller detail in my discussion in Chapter 2 of the "Transportation Tort law School." Leveson defined "safety" in an absolute sense: Safety is freedom from accidents or losses. Readers will later note the similarity with Willie Hammer's 1972 definition. She also cited William Lowrance's view that there is no such thing as absolute safety, and therefore safety should be defined in terms of acceptable loss.   The attendant questions raised by such a definition includes: What is meant by "acceptable" risk? To whom is the risk posed? By whom is it judged to be acceptable? A condition that is acceptable to an employer may not be acceptable to the employee and vice versa.   The questions lead to endless arguments about what level or type of loss is "acceptable?"   Leveson envisions safety, like other qualities, along a continuum, with the left end being freedom from losses and the continuum stretching toward increasing loss. "It is simplest to put safety at the left of the continuum and then determine how close one comes to that ideal." If "safe" is not at the left end of the continuum, then where should it be? [33]

The terrifying air accidents of *ValuJet Flight 592* (May 1996) and *TWA Flight 800* (August 1996) underscored the need to draw a distinction between regulatory compliance for "certification" and "safety" when communicating risks to the public. In discussing the issue of aviation safety risk communication, it is clear that *safety is more than the absence of accidents.*   What then is "safety?" and how should aviation safety information be communicated to the travelling public?   This essay offers a definition of "safety:"

> Safety is the goal of transforming the levels of risk that inheres in all human activity.

The 1997 textbook, *System Safety Engineering and Risk Management: A Practical Approach* is designed to show engineers how to design and build equipment that is safe.   The author, Nicholas J. Bahr, has cited the startling results of an American Society of Mechanical Engineers National Survey (Main & Ward 1992) that found that although most design engineers were aware of the importance of safety and product liability in designs, most did not know how to use the available System Safety tools. Almost 80 percent of the engineers had never taken a safety course in college, and more than 60 percent had never taken a short course in safety through work.   Also, 80 percent had never attended a safety conference and 70 percent had never attended a safety lecture. Bahr then asks a rhetorical question: How do engineers (who are often called to testify in court about design failures) build, and operate systems safely if they have never been prepared for it?[34]   The implications of data on design related aviation accidents as well as the results of my own efforts to obtain information from the National Science Foundation on the number of doctoral dissertations in design engineering are discussed.

Classifying the Literature

In classifying[35] the books on safety, the focus is on the historical setting in U.S. transportation history.   Scope limitations preclude the inclusion of many more journal articles. For example, there are, *Reliability Engineering and System Safety Journal, Hazard Prevention,* the Journal of The System Safety Society and others. Their inclusion would have resulted in an overly ambitious project that could easily exceed 1,500 pages--defeating the original purpose and intent of this essay.   However, despite scope limitations, where relevant, reference will be made to significant technological advances and practices that have impacted safety.

A theme of this book is *to examine the genesis of emerged patterns in safety thinking.*   The plan is to examine the apparently inwardly regenerative and evolutionary spiral of safety thinking that has taken us to where we are to day, at the dawn of the 21st century; in thinking about and understanding "safety." In so doing, we may be best able to meet the new challenges that lay ahead. In the process, we have examined the role of safety activists, (fanatics to many), who dared to challenge the conventional wisdom.   If indeed, "safety is everyone's concern," then safety is too important a subject to be left to the exclusive domain of design engineers and risk managers with their models of economic rationality, or for that matter, lethargic governmental bureaucracies.

Aptly, for our focus on transportation safety, the basic learning structure begins with congressional passage of the *Railroad Safety Appliance Act,* March 2,1893, progressing to the present era of commercial space transportation. This effort located *key* scholarly writings by date of publication. The beginnings of three periods are identified, for purposes of exposition, as follows:

1893   Thesis   Transportation Tort Law School
                         (Safety Engineering Design Emphasis)

1943   Antithesis   Reliability Engineering School
                         (Hardware Reliability Emphasis)

1962   Synthesis   System Safety Engineering School
                         (Hardware, Software Engineering,
                         Human Factors & Organizational
                         Emphasis)

To be certain, the classification scheme itself presents us with several challenges. First, the construct seems to suggest that concern for the identified areas began and ended at neat intervals. This is not the case. There are some overlaps that do occasionally conflict with the order of theoretical presentation. This conflict is most evident when discussing both the Probabilistic Risk Assessment, the Human Risk Assessment approaches; as well as organizational management and administrative vulnerability issues first brought to attention by Barry Turner in 1978. However, on balance, as we examine the various intellectual emphases within the safety discipline, what has emerged is a distinct and visible unifying theme of an apparent self-reinforcing evolutionary spiral toward a comprehensive pattern in thinking about safety. An integral conflict with the classification scheme may be the impression conveyed that "System Safety" is the final "evolution" in the identified patterns in safety thought. That is not the case.

To match the needs of aviation safety practitioners, the majority of whom have a practical "hands-on" background and tradition of vocationalism, the challenge has been to first guide readers, then highlight (ever so briefly) the viable ideas drawn from the vast and often esoteric literature, as well as from relevant newspaper accounts. Where appropriate, chapters contain concluding remarks of practical recommendations, admonitions and "Best Practice" drawn from the literature. However, it should be emphasized that this essay is not designed to be a "training manual" or cookbook. The essay is designed to educate aviation safety practitioners and engineering students on the broad outlines of the history, theory and practice of "safety management" in the United States. The hope is that it can serve as a concise guide and reference point for an understanding of the evolution of safety thought and actions by private citizens, as well as to stimulate inter-disciplinary thinking about transportation safety and consideration of the issues raised.

Outline of the Book

Chapter 2 extensively chronicles the activities and accomplishments of Lorenzo Coffin, who in the last quarter of the 19th century, beginning in 1874, carried on a crusade almost single-handedly to force railroads to adopt basic safety devices to reduce the carnage to railroad personnel. In 1893 congress responded and passed *The Railroad Safety Appliance Act.*

The Act marked the origin and development of legal and federal regulatory actions in transportation safety, hence the "Tort Law School." This Chapter sets forth the moral, legal and theoretical framework for modern-day aviation safety advocates such as Ralph Nader, Mary Schiavo, John King, and Captain "X" who are seen as Coffin's spiritual heirs. It supports the thesis that *safety is everyone's concern*. We briefly examine select landmark auto and aviation safety related lawsuits including, *Grimshaw v. Ford Motor Company (1978), Volkswagen of America, Inc., v. Germaine Gibbs (1989), United States v. Edward Upton et. al. (1994) (Eastern Airlines), and Hope et. al. v. McDonnell Douglas   et. al.,* that have shaped the "Tort Law School."

Chapter 3 is devoted to the "Reliability Engineering School" and its traditions of understanding "failure," "fault," "failure condition," and a very brief overview of some of the analytical tools of the Reliability Engineer. The natural progression to the new branch of probabilistic risk assessment that has been embraced by the nuclear and other high-risk technologies is also briefly examined. The further evolution to human reliability analysis is also discussed.

The fundamental premise of the "Reliability Engineering School" is that *equipment reliability and safety are synonymous*. The issue of redundancy and safety is of particular interest. The pioneering work of the U. K's Barry Turner and Nick Pidgeon on the complex relationship between organizational culture and safety is examined. The work of the "High Reliability" organizational theorists from the University of California, Berkeley and the "Normal Accidents" organizational theorists proponents led by Perrow of Yale University and Sagan of Stanford University are examined in the context of the aviation system capacity problem and the issue of risk communication to the public. This chapter has also drawn from relevant published newspaper reports as well as publicly available information on the World Wide Web. The phenomena of software-intensive systems have presented new challenges for aviation safety and efficiency.

In Chapter 4, we briefly explore the representative literature of the "System Safety Engineering School" and its evident evolutionary spiral from the "Reliability Engineering School." It is by no means the intention of this Chapter to convey the impression that "System Safety" is the final "evolution" in the identified patterns in safety thought. However, "System Safety" does aspire to synthesize and blend traditional engineering and management science/organizational risk issues with human factors analyses. It is beyond the scope of this essay to assess and evaluate its success.

The intent of this Chapter is not to duplicate the voluminous material on "System Safety," but rather to provide concise educational reference material in a framework designed to enhance critical thinking about the safety challenges of the 21st century. In this Chapter, we briefly examine the Roland & Moriarty classic and other more recent textbooks texts designed to be the System Safety Engineer's reference manual. Appropriate human factors literature not covered in Chapter 3 is also reviewed. Issues of safety education and training first raised by Stephenson are also discussed. This Chapter also includes material on a "real-world" safety risk management process for aviation.

Chapter 5, Conclusion, issues a call to the President's Science and Technology Advisor and the National Academy of Sciences to investigate and reassess the historical rigidity of our major engineering schools in not addressing "safety education." Educational assessments of key professions are nothing new. In drawing from the literature and "best practice" the chapter also includes an ideal summary combination of experience, training, education and knowledge that the regulatory mandated, but undefined, position of "Airline Safety Director" should possess.

Notes

[1] Views expressed to Mr. Terry Kelly, Safety Manager, NAVCANADA and Mr Steven D Smith, System Safety/Risk Management Instructor, Federal Aviation Administration, by trainees at the 1998 NAVCANADA Aviation Safety Management Training Course
[2] U.S. Department of Transportation, Bureau of Transportation Statistics, *Transportation Statistics Annual Report*, 1997, "U.S.GDP Attributed To Related Final Demand," Chapter 2, Table 2-1A.
[3] *See* Keith M Henderson, *Emerging Synthesis in American Public Administration (Asia Publishing House, 1966)* That effort successfully focused attention upon the representative scholarly works that are relevant to the discipline of American Public Administration as a whole, rather than the more limited areas of Administrative Law, Management, Business Administration, and Political Science. As a "practical study," American Public Administration in the past has borrowed liberally from numerous fields including Mathematics, Social Psychology, Economics, Sociology, Cultural Anthropology and Psychology.
[4] The entire paragraph is paraphrased from Martin L. Shooman, *Probabilistic Reliability An Engineering Approach* (New York: McGraw-Hill, Inc., 1968), pp. X, 1.
[5] This paragraph is adapted from Harold E. Roland & Brian Moriarty, *System Safety Engineering And Management* (New York: John Wiley & Sons, Inc., 1990). 2nd ed., pp. 8-11.

[6] Ralph Nader, *Unsafe At Any Speed,* (New York: Grossman Publishers, Inc., 1965).

[7] Captain Vernon W. Lowell, *Airline Safety Is A Myth,* (Bartholomew House, 1967)

[8] John Godson, *Unsafe at Any Height*, (New York: Simon and Schuster, 1970), pp.29-30.

[9] Brenda McCall, *Safety First at Last.* (New York: Vantage Press, 1975).

[10] William E.Tarrants, *The Measurement of Safety Performance,* (New York: Garland Publishing, Inc. 1980, *Preface*

[11] H.W. Heinrich, Dan Petersen, Nestor Roos, *Industrial Accident Prevention,* (New York: McGraw-Hill, 1980) 5th ed., p. 21.

[12] IBID., pp. 357-367.

[13] John V. Grimaldi and Rollin H. Simonds, *Safety Management* (Homewood, IL: Richard D. Irwin, Inc., 1975) 3'd ed., pp 39-55.

[14] Willie Hammer, *Occupational Safety Management And Engineering,* (Englewood Cliffs, NJ: Prentice-Hall, Inc.1981), 2nd ed., p.2.

[15] Kenneth E.F. Watt, The Titanic Effect, (New York: E.P. Dutton & Co., Inc., 1974), pp. 6-7.

[16] Professor Lee Clarke, Department of Sociology, Rutgers University: "The Disqualification Heuristic: When Do Organizations Misperceive Risk?" *Sandia Report.* Proceedings of the High Consequence Operations Safety Symposium. July 12-14, 1994. Organizational Strategy and Management Session.

[17] Barry Turner, *Man-Made Disasters,* (London: Wykeham Publications, 1978).

[18] Nick Pidgeon, *"The Limits To Safety? Culture, Politics, Learning and Man-Made Disasters,"* Draft paper prepared for, Special Issue of *The Journal of Contingencies and Crisis Management,* Vol. 5, Number 1, March, 1997.

[19] Nick Pidgeon, "Systems, Organizational Learning, and Man-Made Disasters," *Conference Proceedings, Probabilistic Safety Assessment And Management (PSAM 4),* Vol. 4. p. 2687. September 1998.

[20] IBID., p. 2688.

[21] Barry Turner and Nick Pidgeon, *Man-Made Disasters,* (Oxford: Butterworth-Heinemann, 1997).

[22] Charles Perrow, *Normal Accidents,* (New York: Basic Books, Inc., Publishers, 1984), p.11.

[23] IBID. p. 4.

[24] IBID. p. 146.

[25] James Reason, *Human Error,* (Cambridge: Cambridge University Press, 1990), p.1.

[26] James Reason, *Managing The Risks of Organizational Accidents,* (Brookfield, VT: Ashgate Publishing Co, 1997).

[27] Steven M. Casey, *Set Phasers On Stun And Other True Tales of Design, Technology and Human Error.* (Santa Barbara, CA: Agean Publishing Co., 1998), 2nd ed., p.11.

[28] Scott D. Sagan, *The Limits of Safety: Organizations, Accidents and Nuclear Weapons,* (Princeton, NJ: Princeton University Press, 1993), p.19.

[29] Nancy Leveson, *Safeware System Safety and Computers* (Reading, MA: Addison-Wesley Publishing Co. 1995), p. vii.

[30] Felix Redmill, Morris Chudleigh and James Catmur, *System Safety. HAZOP and Software HAZOP,* (West Sussex, England: John Wiley & Sons Ltd., 1999).

[31] IBID., p. 17.

[32] IBID., p. 532.

See Also: *Software System Safety Handbook.* Produced by the Joint Software System Safety Committee, September 30, 1997.

Note: On February 21st, 1990, Public Television's *MACNEIL/LEHRER NEWSHOUR* featured a story on "Software Safety," Show #3673. This was during the development of RTCA DO-178B: *Software Considerations In Airborne Systems And Equipment Certification.* The author served as the Designated Federal Representative on the international committee.

The differentiating software levels are of special interest to the issue of measuring how dangerous is safe enough? and what price safety? For example, Annex A of DO-178B presents some 59 objectives to be met during the development process for Level B or Level C software. Some 42 of these objectives are the same for Level B and Level C. Of the remaining 17 objectives, 6 are expressly required for Level B; 11 are identical for both Levels with the exception that Level B objectives require independent verification.

[33] Leveson, Op. Cit., p.181.

[34] Paraphrased from Nicholas J. Bahr, *System Safety Engineering and Risk Assessment: A Practical Approach,* (Washington, DC: Taylor & Francis, 1997), p.12.

[35] See Elmer Sprague & Paul W. Taylor, *Knowledge and Value,* (New York: Harcourt, Brace & World, Inc., 1959), p. 684. This classification framework of the safety literature into thesis, antithesis, and synthesis has revealed an evolutionary spiral by which safety thinking has come into and gone out of existence.

This exposition owes much to the philosophy of G.W.F. Hegel (1770-1831), German philosopher and historical theorist. The framework is based on the Hegelian dialectic. Described briefly, Hegel views history as a conflict between Spirit, which is trying to impose its form (idea) on the World, and the World (matter), that resists the impositions of Spirit. Spirit is bound to succeed finally, but its success can only be achieved little by little. Spirit clashes with the World and imposes a little of its form on the World.

This success encourages Spirit to a new clash, and the world becomes a little more spiritualized. This clash of the antitheses, Spirit and the World, is the most important example of Hegel's dialectic. *Whatever exists is the product of the clash of opposites.* However, the product of these clashing opposites is short-lived; for the new entity, which their union creates calls forth its own opposite with which it must clash in order to continue the dialectical growth of the World. For Hegel, if we could understand the form that Spirit ultimately wishes to impose on the World, we should know the goal of history.

Throughout much of the 19th and 20th century, although Hegel's ideas became outmoded, none appeared to have escaped his teaching. Karl Marx was enormously impressed by Hegel's philosophy of history, but he subjected it to what might be called a dialectical transformation, or as he put it himself, he stood Hegel's philosophy on its head. Marx rejected the idea that a supramundane spirit, or intelligence, is endeavoring to transform the world. The only intelligence to be reckoned with is that of the minds of individual human beings and that the individual's mind is conditioned by the world in which he lives. The material conditions, or more exactly the economic arrangements, of the world shape the minds of the people who live under them.

2 Transportation Tort Law School

"I discovered it was taken as a matter of course that railroad men of necessity be maimed and killed."

Lorenzo Coffin

Spirited Moral Outrage

The *Railroad Safety Appliance Act, 1893,* is the starting point in the emergence of the "Transportation Tort Law School" - the historical origin of patterns in transportation safety thinking. "Torts" have been defined as a comprehensive legal term for any type of civil wrong, with the exception of a breach of contract, for which the courts will provide relief in the form of damages for injury suffered. The three basic goals of the law of torts are: (1) to compensate individuals who have suffered a loss or injury as a result of another's action or conduct; (2) to force only the wrongdoer to be responsible for the costs; and (3) to prevent future loss or injury.[1] This chapter includes selected references to highly publicized American tort law cases in the automobile and air transportation sectors. These cases have set hitherto unseen multi-billion dollar jury awards in California and new legal precedents in Florida of criminal liability and murder indictments in an accidental airline crash. The chapter also references the role of single-minded individuals, fanatics to many, who have brought about changes in safety thinking.

Lorenzo Coffin's legacy of regulation as an early 19th century advocate of rail transportation safety and the doctrine that "safety is everyone's business," as well as his work in drafting the *Railroad Safety Appliance Act, 1893,* has earned him an immortal place in U.S. transportation safety history. Coffin's historic efforts resulted in the tradition of governmental intervention to regulate the safety of transportation. The following material on Coffin is paraphrased from Stewart H. Holbrook research.[2] Coffin's work deserves a full appreciation since it was destined to serve as a model for Ralph Nader, the 20th century's

23

celebrated crusader for automobile, highway and air transportation safety. This chapter also documents the emergence of Ralph Nader as the most famous advocate of the "Tort Law School" of transportation safety. It would take another 70 years before the tort law liability actions of Ralph Nader would spur government regulatory actions to effect automobile transportation safety, which had become the dominant mode of inter-city travel.

The Mind of A Safety Fanatic

To understand the beginnings of the early concern for transportation safety, we are introduced to Lorenzo Coffin, the New Hampshire born, successful Iowa farmer, whose life at the age of 51 in 1874 took a dramatic new turn. In that year, while riding a passenger-freight train, he witnessed an accident to a railroad brakeman who was switching freight cars. The worker lost the two remaining fingers of his right hand. He had lost the others in a similar accident the year before. Coffin learned that few-experienced brakemen possessed all of their fingers; they were lost in the dangerous operation of coupling cars by the pins then in use. The workers had to stand between the cars being coupled, in order to direct the pin into the socket. This took a lot of fingers; many whole hands, and often lives were lost in applying brakes in adverse weather or other environmental conditions.

The accident he had seen and the knowledge he had gained produced a powerful effect on his thinking about safety. He knew that an automatic coupler, that precluded the brakeman standing between moving cars, had been invented and was being manufactured and that George Westinghouse had patented his air brake that was even then being used on a few Eastern lines. Increasingly, Coffin rode about the Middle West on freight cars, always talking with the crews. They told him that the new brake and the new coupler were not being widely adopted because of the cost. *He vowed there and then that he was going to do something about it, that he had found his real life's work.* And so, in the words of Stewart Holbrook, he "became a fanatic, a monomaniac." Years later, Coffin said:

> My first job was to arouse the public to this awful wrong, this butchering of these faithful men who were serving the people at such fearful risk of life and limb. Why, I discovered that it was taken as a matter of course that railroad men of necessity be maimed and killed.

Railroad travel in the 1870's was a precarious affair. The safety of passengers was hardly considered. From about 1850 until the end of the century, wrecks caused by head-on collisions, by derailments, by falling bridges and trestles continued to be a national scandal. In 1881, there were some 30,000 casualties, either killed or maimed for life because of accidents with hand brakes and coupling pins.

With few exceptions, the railroads were resistant to Coffin's suggestions for the use of air brakes and automatic couplers; and they feared nobody. "They were buying the influence of United States senators and congressmen, just as they bought rolling stock." They operated trains when and where they pleased. They charged what they would for freight and passengers. In American railroads, the policy of laissez faire was to be seen in its entire flower. By 1880 American railroads were probably the most arrogant corporations in the country; and Coffin set out to tackle these giants single-handed.   He became known as the "Airbrake Fanatic." Railroad officials got to know him either as a bore or a gadfly. Snubbed, often thrown bodily out of offices, he turned to the press and was promptly ignored.   The only outlets he could find for his protests were the small religious, family, and farm periodicals where he poured his discontent with things as they were.

In direct and appealing language, often with Scriptural allusions, he engaged our emotions and attacked the reasoning of the railroads. He described in detail, dripping with gore, the accidents, and mentioned the tragedies brought to families by loss of their breadwinners who had been killed or maimed because of accidents with hand brakes and coupling pins. The butchery was unnecessary because George Westinghouse had invented a brake that would do away with most of this murder, but that because of the cost the railroads, "sweating golden dividends at every pore," would not install it.

In 1883, when he had passed the age of sixty, his crusade showed its first small results.   He was made railroad commissioner of Iowa. Immediately, he wrote to railroad officials, to newspapers, labor unions, societies, and individual men and women and asked for moral support to get the railroads to adopt the simple technology that would greatly reduce injury and fatality to railroad employees. He invited himself to conventions of railroad officials.   Once there he stood up and accused the railroads of committing mayhem and murder.   That was ineffective and he decided to try a new approach—*safety engineering design. (My addition).*

The Role of Engineering Design & Testing

Coffin managed to get the Master Car Builders Association to agree to a test of airbrakes on a long freight train. The first test was held in 1886 and was disappointing. The second trials, attended by George Westinghouse, were held in May 1887. The tests failed once more, and again he was denounced as a wild-eyed fanatic. However, when Westinghouse returned to Pittsburgh, he personally went to work in his shops to perfect an air brake that would stop a fifty-car train. *This historical event marked the first attempt in transportation to merge the science of engineering design with the moral and spiritual concern for the safety of personnel and ultimately passengers. (My commentary).*

Late in the summer of 1887, the third and what have since been famous in railroad history as the Burlington Trials were held on a long grade of track of the Chicago, Burlington & Quincy, eight miles west of Burlington, Iowa. Both Westinghouse and Coffin were present. The two watched, said an eyewitness, while "the immense train was hurled down the steep grade at forty miles an hour." At a signal the air brakes were applied, and "the train came to a standstill within five hundred feet and with hardly a jar." Lorenzo Coffin, aged and weather-beaten, stood close to the track and bystanders saw tears of joy stream from his eyes and run shamelessly down his lined leathery face. "I am," he cried with great emotion, "the happiest man in all creation!" It was an epochal day in American railroad history. Not before had a long heavy train of cars been stopped by air, quickly and without harm to men or equipment. Coffin, still naïve in respect to railroads of the time, thought his work was done. It was really just beginning, for the railroads were not anxious to adopt the new device.

A Presidential Pen

The time-honored presidential tradition of providing souvenir pens after the signing of legislation was again witnessed. After the successful tests, Coffin, the safety fanatic, had drafted the first railroad safety-appliance law ever written. It required that all trains operating in Iowa should be equipped with air brakes and automatic couplers. The act was made into state law. The railroads disregarded it. Yet its passage and the flouting of it brought national attention to Coffin and his safety crusade. Then, in 1888, the Interstate Commerce Commission came into being. One of its first acts was to invite the state railroad commissioners of the nation to a meeting in Washington to discuss various traffic rate problems. Although he was by then no longer a railroad commissioner, he planned to attend the meeting. First he rewrote his safety-appliance law to fit the country as a whole; then he went to Washington.

As soon as the gathering was called to order, ready to discuss rates, he stood up, "straight and tall despite his years" and told the assembled experts, before they could stop him, more about railroad accidents than they had ever guessed, or wanted to hear. He piled horror upon horror. "These are cold facts," he said, his voice shaking. He painted scenes that made his listeners squirm, and left many of them, as they were later to admit, wet-eyed. He was eloquent and he knew what he was talking about. His address had a lasting effect on the audience, including the Interstate Commerce Commissioners themselves. However, the meeting refused to approve Coffin's proposed national act relating to safety appliances. The time still was not quite ready. Coffin went back to his farm, not in defeat, but to prepare for the next attack. He raised money from his own resources, then returned to Washington, to stay for four years, leading a Spartan life in cheap boarding houses. He *drafted a law* that if passed would cover every mile of every railroad in every state. The law was to apply to all railroads alike, and it would impose on them what Coffin said was the absolute necessity of air brakes and automatic couplers. He lobbied senators and congressional representatives. He rewrote his proposed law again and again until he managed to get a bill through the House and Senate:

> It shall be unlawful for any common carrier engaged in interstate commerce by railroad to use on its line and locomotive engine in moving interstate traffic not equipped with a power driving-wheel brake and appliances for operating the train-brake system, or to run any train in such traffic that has not a sufficient number of cars in it to so equipped with power or train brakes that the engineer on the locomotive drawing such train can control its speed without requiring brakemen to use the common hand brake for that purpose.

Finally, on March 2, 1893, often called the greatest day in railroad history, President Benjamin Harrison signed the *Railroad Safety Appliance Act,* and gave Coffin the pen with which he signed the bill into law.

Discovering New Safety Issues

The *Railroad Safety-Appliance Act* brought immediate and striking improvement, reducing the accident rate to employees by more than fifty per cent, while the passenger accident rate, long a scandal, fell to almost nothing. The act had another influence. Once the more progressive roads discovered *the positive effects of safety* on their business such as an increase in railroad running speeds, improvements in safety increased. The knowledge that the safety of their employees, as well as the safety of passengers, paid dividends, marked a milestone in safety thinking.

Coffin lived well into the 20th century (he died, January 15, 1915, at the age of ninety-two) and witnessed the steady improvement and safety of train travel. However, his main work was not done yet, "being a fanatic he still must have something to live for." This turned out to be the problem of alcoholism among railroad employees. He was convinced that alcoholism was second only to hand brakes and link-and-pin as a cause of accident and disaster. Although there were rules prohibiting the use of intoxicants while on duty, a large number of railroad workers circumvented them to the detriment of the safety of passengers and workers alike. His answer to the problem of alcoholism among railroad workers was the Railroad Temperance Association, which he founded at his own expense. With missionary zeal, he distributed more than 250,000 little white buttons to trainmen and other employees who "had given their word to abstain…(during)…performance of duties." He also helped to found in Chicago the Home for Aged and Disabled Railroad Men. In 1907 he campaigned for the governorship of Iowa on the prohibition ticket.

In 1946, in *The Story of American Railroads,* the author, Stewart H. Holbrook concluded the Chapter: "The Airbrake Fanatic," by stating that Coffin was a "true fanatic to the last, one of the most useful fanatics this country has produced." No better example of a monomaniac who was of great benefit, could be found than Lorenzo Coffin, the bearded and forgotten fanatic whose one track mind and grim determination were of incalculable aid in bringing to success the two greatest railroad inventions since the steam locomotive: George Westinghouse's air brake and Eli Janney's automatic coupler. Holbrook lamented the irony that Americans, (up to that time), would erect monuments to military heroes but few knew that Lorenzo Coffin, a saver of countless lives, ever lived.

Modern Day Transportation Safety Advocates

A century after Coffin's activities, through our perspective of modern lenses, we see Coffin's legacy of railroad safety advocacy through a combination of spirited moral outrage, the incorporation of fortuitous safety-related research and development and sound engineering design principles in modern transportation. This was the earliest manifestation of the opening of the safety discipline to persons of different backgrounds in the belief that: *safety is too important to be left in the hands of any one group.* It is a theme that would be reinforced in the last half of the 20th century with the dominance of automobile and air transportation and the safety activities of Ralph Nader.

To maintain the chronological order of the presentation of the material, brief mention is made of a technological innovation that proved the axiom that safety is everyone's concern—the invention of the traffic light and the safety hood, later known as the gas mask by Garrett A. Morgan (1875-1963). On November 20, 1923, Morgan, an African-American entrepreneur, was awarded U.S. Patent No. 1,475,024 for his invention of a three-way automatic traffic signal for which he received an U.S. government citation.[3] In 1997, the U.S. Transportation Secretary, Rodney Slater, unveiled the department's Garrett A. Morgan Technology and Transportation Futures Program, a national education program for students.[4]

Auto and Truck Safety

It was inevitable that the initial thread of concern with transportation safety, first with rail, would naturally spread with the growth of the automobile and later aircraft transportation. True to the American muckraking tradition of the 19th century, in many respects the activities of modern-day safety advocates such as Ralph Nader and Mary Schiavo, would appear to be Coffin's spiritual heirs. In fact, in the preface to *Unsafe at any Speed* (1965) Ralph Nader quotes the 19th century editors of *Harper's:* "So long as brakes cost more than trainmen, we may expect the present sacrificial method of car-coupling to be continued."[5] Nader observed that injured trainmen did cause the railroads some "operating dislocations." However, highway victims cost the automobile companies "next to nothing" and the companies are not obliged to make use of safety-related research and development. For Nader, the lack of public scrutiny of the automobile was a function of the lack of publicly available information. He then went on to explain why the automobile had remained the only transportation vehicle to escape being called to full accountability.

It is impossible in this book to evaluate the thousands of design-related automobile safety lawsuits that have become enshrined in U.S. tort law, the history of which is better left to legal historians. However, passing reference is made to major lawsuits that support the existence of a distinct "Transportation Tort Law School" of safety. Nor would those same page limitations permit a thorough report on Ralph Nader's over three-decade efforts on behalf of consumer transportation safety beginning in 1965 with the publication of *Unsafe At Any Speed*. That book documented the tragedy of the General Motors' Corvair in the annals of U.S. automobile history. General Motors responded by monitoring him with private detectives. He successfully sued General Motors and invested in a Center for Responsive Law, which served as a siren for the dawning of the consumer movement.

The first Corvair automobile was introduced on 13 September 1959 with subsequent models produced through 1965. By October 1965, more than one hundred lawsuits alleging instability in the Corvair had been filed around the country. In the summer of 1965, three of them were decided in court. General Motors denied the charges, and instead blamed the accidents on driver negligence. In none of the three suits did General Motors reveal the technical data and test results that would have placed before the public the full facts about the Corvair.[6]

Ralph Nader stirred the nation by exposing acts of industrial irresponsibility in the conception, design and development of the Corvair. As a study in risk management, it provides a tragic example of the *transferring of risk*. The risks were transferred to the drivers of the car by requiring them to have and maintain fairly esoteric knowledge so as to cope with the vehicle's inherent directional instability. This unsafe procedure became "a dangerous situation to the extent that the vehicle is sensitive to oversteer or directional instability from variations in tire pressure."[7]

The book provides a case study in the evolution of the consumer movement and validation of the axiom that safety is too important to be left to designers and professional risk managers. It also provides illustrative examples of the behavior of organizations toward those "outsiders" and "self-styled experts" who dare to pose questions. The role of government is critically examined. For Nader, as the gap between existing design and attainable safety rises, so do the moral imperatives to dedicate engineering and investment energies to use new approaches that will make the automobile responsive to the safety requirements of motorists.   He recognized that "regulation of the automobile" must go through three stages: (1) public awareness, (2) legislation, and (3) continuing administration or oversight.   Since auto safety ideally should reflect advances in research and development, administrators have a responsibility to advance it.   Without full disclosure, congressional oversight and "participation by a consumer-oriented constituency" of professionally qualified citizens, obsolescence and bureaucratic inertia will "stifle the purpose of even a properly drafted law." The book concludes with a call for a crusade by the many people in the automobile industry who know both the technical capability and appreciate the moral imperatives.   But their timidity and conformity to the rigidities of the corporate bureaucracies have prevailed.   However, when and if the automobile is designed to free millions from unnecessary mutilation, they, like their counterparts in universities and government who know of the suppression of safer automobile research and development yet remained silent, will look back with shame on the time when common candor was considered courage.[8]

An updated version of *Unsafe At Any Speed* (1972), documented the early years of the National Highway Traffic Safety Administration (NHTSA).   The book also documented how the "uncooperative, obstructive, and noncompetitive posture of the auto industry helped generate support within the Department of Transportation and congress for strengthening the 1966 auto safety legislation." Nader established a Clearinghouse for Professional Responsibility to encourage "people of conscience, struggling against corporate and bureaucratic pressure in their organizations, to get suppressed safety information to the public."[9] Nationwide, highway deaths average about 45,000 annually.

Despite the legislative and regulatory actions that followed the Corvair, less than a decade later, it would appear that the Ford Motor Company, in designing the Pinto, had not fully anticipated the intensity of consumerism and society's views on safety.   The deliberate *transfer of risk* by corporations after their cost benefit calculations would again rivet the nation's imagination.   In the landmark case, *Grimshaw v. Ford Motor Company,* (1981), the Court of Appeal of California let stand a judgment against Ford. The case involved a 1972 Ford Pinto hatchback automobile. The car unexpectedly stalled on a freeway, erupting into flames when it was rear ended by a car proceeding in the same direction. Mrs. Lilly Gray, the driver of the Pinto, suffered fatal burns and 13-year-old Richard

Grimshaw, a passenger in the Pinto, suffered severe and permanently disfiguring burns on his face and entire body. Grimshaw and the heirs of Mrs. Gray sued Ford Motor Company. Following a six-month jury trial, verdicts were returned against Ford. Grimshaw was awarded $2,516,000 compensatory damages and $125 million punitive damages; the Grays were awarded $559,680 in compensatory damages.[10] The design of the Pinto was such that upon impact, its gas tank would be pushed forward and caused it to be punctured by the flange or one of the bolts on the differential housing so that fuel would be sprayed from the punctured tank and enter the passenger compartment. Harley Copp, a former Ford engineer and executive in charge of experimental crash testing testified that the highest level of Ford knew this. Ford management made the decision to go forward with the production of the Pinto, knowing that the gas tank was vulnerable to puncture and rupture at low rear impact speeds creating a significant risk of death or injury from fire, and that "fixes" were feasible at nominal cost.

As mentioned earlier, Perrow cited the case of the Ford Motor Company decision not to buffer the fuel tank in the Pinto, and at the General Motors Company when it rejected warnings from engineers that the Corvair would flip over for the lack of a $15 stabilizing bar. His criticism of the risk management discipline is that it acknowledges the difference between voluntary risks such as skiing and hang-gliding, and involuntary ones such as leaching of chemical wastes. It does not acknowledge the difference between the *imposition of risks by profit-making firms* who could reduce that risk and the *acceptance* of risk by the public where private pleasures are involved (skiing) or some control can be exercised (driving). "All are bundled up in a vague reference to market principles." [11]

Grimshaw v. Ford was a landmark in transportation safety law from the view that it introduced the notion of "corporate criminal liability for homicide into industry decision making which operates to promote product safety."[12] It marked a shift from a *regulatory* approach to a *liability* approach and the full-scale emergence of the "Tort Law School" and Ralph Nader as its pre-eminent voice. During this period, others proposed the imposition of certain tenets for design engineer responsibility. Among these was the need to design and test in accordance with the highest professional standards. Accordingly, engineers should refuse to sign reports leading to faulty or unsafe products; to adhere to the Code of Ethics statement that responsibility to the public is paramount; and to report problems internally, notifying appropriate authorities or agencies if corrections are not made.[13]

During the 1980s, lawsuits involving the Audi 5000 also provided the transportation "Tort Law School" with documented material on the problem of liability for accidents in which the car allegedly experienced sudden, unintended acceleration. A problem, which the car manufacturer said, had been caused by "driver error." A number of lawsuits were settled out of court. Despite massive recalls, NHTSA concluded that no mechanical defect caused the car to suddenly and unintentionally accelerate. The NHTSA report said that design features allowed drivers to inadvertently depress the wrong foot pedals, causing the cars to lunge out of control. Ralph Nader's Center for Auto Safety expressed disappointment.[14]   One lawsuit reached the U.S. Supreme Court, *Volkswagen of America Inc. vs. Gibbs.* The Court let stand a $100,000 punitive damage award against the parent company of the Audi, resulting from a 1983 accident in which an Audi 5000, imported by Volkswagen, smashed into an apartment. Damages were awarded to the occupant of the apartment. A New Jersey Court jury said that the accident was caused by sudden acceleration because of the defective design of the gas and brake pedals.[15]

The 1990s witnessed the actions of a self-described "ordinary homemaker," Janette Fennell, in crusading for the auto industry to install a safety feature long missing in cars—an internal emergency trunk release. The Trunk Releases Urgently Needed Coalition began after Fennel and her husband had been locked in the trunk of a Lexus sedan and left for dead in October 1995. They survived after struggling for "one hour and forty-five minutes of sheer terror" to find the internal trunk release latch. Fennel then started doing research, calling car companies, the NHTSA, and the Society of Automotive Engineers. She learned that many cars did not have an internal trunk release, let alone one that was easy to find and operate. There was no NHTSA standard requiring such a device. There were no NHTSA records on incidences and consequences of trunk entrapment, forced or accidental.   Initially, she was not taken seriously, *an understandable neurotic.* However, in the summer of 1998, after 11 children died from being inadvertently trapped in the trunks of motor vehicles, Fennell received early attention and commitment from Ford Motor Co., and a groundswell of support. Fennel's coalition and the National Safe Kids Campaign presented their findings to the NHTSA-appointed ad hoc Expert Panel on Trunk Entrapment. They reported that since 1970, an estimated 1,175 people have been victims of trunk entrapment in 992 separate incidents. As a result, without regulation, the auto industry decided that acting on trunk safety "was the right thing to do."[16]

In August 1999, for the first time in 60 years, the Federal Highway Administration (FHWA) announced plans to propose new limits on the number of hours truck drivers are allowed to work. Safety groups and industry officials have been battling for years over how much control the government should have over truckers' hours. The FHWA is expected to force drivers to be off duty for at least 14 hours of a 24-hour period. Currently, drivers have to be off duty for only eight hours a day, although actual driving time is limited to 10 hours. Proponents of the limits argue that if truckers are off for only eight hours, they don't spend it all sleeping. The proposal will make it more likely that drivers will sleep eight of the fourteen hours. According to preliminary government statistics, about 5,300 people died in truck-related crashes in 1998. Less than 4 percent of those crashes were caused by truck drivers' drowsiness. Predictably, the American Trucking Association is opposed to the plan and has petitioned the FHWA to reveal the scientific evidence that supports the plan. The trucking association said that limiting work hours would not make roads safer and it is more important that drivers have regular schedules with enough hours of sleep.[17]

In 1999, the trucking sector also witnessed a major safety innovation brought about by the activism of an accident victim's relatives. The parents of Ben Cooper, a high school senior who died in August 1997 when a loaded dump truck overturned on his car, collected a $4.6 million settlement from a trucking firm and the excavation company that loaded the truck. Cooper's high-profile parents—Richard M. Cooper is an attorney, Judith C. Areen is dean of the Georgetown University Law Center—spent two years trying to wring some good from their grief: They became *advocates for industry safety standards* in the simple hope of improving public safety. They wrote Washington metropolitan area contractors and excavators, urging company executives to do a better job of monitoring trucking companies, and safe practices through a proposed set of eight recommendations to be added to their contracts.[18]

On Friday, July 9th 1999, a new landmark for product liability was set when a Los Angeles Superior Court jury ordered General Motors to pay a record $4.9 billion in damages. The award damages were to two women, Patricia Anderson and family friend Jo Tigner and Anderson's four children who were trapped and burned when the gas tank of their 1979 Chevrolet Malibu exploded after a rear-end collision on Christmas Eve 1993 by a drunk driver.   Attorneys alleged that GM chose not to spend $8.59 per car to correct the defect.   Internal documents presented in the case stated that gas tanks should be placed no closer than 15 inches to the rear bumper.   Nonetheless, the Malibu's gas tank was 11 inches from the bumper.   Central to the GM case and the jury's record verdict was the 1973 internal GM "value analysis," a memo written by an engineer that calculated the cost to the auto manufacturer of preventing fuel-fed fires. The engineer, Edward Ivey, used an assumption that "each fatality has a value of $200,000." And that there were a maximum of 500 fatalities per year in GM automobiles from fuel fires. The analysis said the deaths in such accidents were costing GM $2.40 per automobile. Ivey also wrote, however, that his analysis must be tempered by the thought that "it is really impossible to put a value on human life." The jury foreman was quoted as saying: "GM has no regard for the people in their cars, and they should be held responsible for it."[19]   On August 27th 1999, the Los Angeles Superior Court Judge Ernest Williams reduced the jury damages award to $1.2 billion.   Judge Williams concluded that GM had placed the Malibu's fuel tank behind the axle to "maximize profits, to the disregard of public safety."   Not unexpectedly, GM plans an appeal.

Air Transportation

Let us now turn our attention to selected lawsuits, books and other publicly available material involving commercial air transportation safety. However, before doing so, it would be important to provide some background information. In 1996 the rare but tragic air accidents involving *ValuJet 292* and *TWA 800* generated intense public criticism of the Federal Aviation Administration's (FAA) approach to safety regulation and management of the National Airspace System. Ironically, the FAA's *1996 Strategic Plan* proudly stated that "FAA talked to ValuJet Airlines to gain business advice from a small, efficient operation in a highly competitive industry." In a 1997 employee survey, a bare majority, 51 percent, of FAA employees believed that the agency has improved aviation safety and 56 percent thought that there was validity to the public criticism of the agency's safety performance.[20]

In the United States, the "Transportation Tort Law School" experienced a major historical breakthrough in Florida.   In an unprecedented case, on July 13th, 1999, Miami-Dade County State Attorney, Katherine Fernandez Rundle, filed murder and manslaughter charges for the fiery May 1996 ValuJet crash in the Everglades that killed 110 people.   "This crash was completely preventable.   It was not an accident like many other crashes. It was a crime." The indictments marked the first time in U.S. history that criminal charges have been brought in an airline accident. Excerpts from various newspaper accounts follow.

The Miami Herald reported that SabreTech, an aircraft maintenance company, bore the brunt of the criminal indictments.   The Miami-Dade State Attorney filed 110 counts of third-degree murder against SabreTech, one for each of the victims.   State prosecutors also charged the company with another 110 counts of manslaughter and one count of unlawful transportation of hazardous materials.   The United States Attorney's office also announced a 24-count federal indictment against SabreTech and three former employees.   The federal charges included conspiracy to make false statements to the FAA, putting short-term monetary interests ahead of public safety; and making false statements and representations and using falsified documents, related to the removal of oxygen generators from ValuJet planes. The three employees had signed off on FAA-approved work orders, vouching that they had placed the required safety caps on the canisters to disarm them.   As a result of the federal charges, SabreTech faces about $6 million in fines and restitution, and the three employees face up to 55 years in prison and more than $2.7 million in fines and penalties.

The National Transportation Safety Board found that the crash was caused by a fire that started in the aircraft's forward cargo compartment, ignited by 144 volatile oxygen-generating canisters that Sabre mechanics had removed from two ValuJet MD-80s. The canisters, which can heat up to 500 degrees, were improperly secured, labeled and packaged, before they were delivered for ValuJet's loading aboard the fatal flight. The canisters had been removed after ValuJet hired SabreTech to refurbish its MD-80s. As part of the retrofitting, mechanics pulled out-of-date oxygen generators from the planes to replace them with new canisters. When they couldn't find protective caps, they cut, tied or taped up the pull strings that ignite the canisters and placed them in boxes.   In testimony obtained during the Safety Board's investigation, the employees admitted that they had not placed the required caps on the canisters. Neither SabreTech nor ValuJet had ordered the yellow caps, which would have cost a total of $9.16, including tax.

In its final report, the NTSB said that ValuJet, SabreTech and the FAA shared responsibility for the fatal crash. It said that ValuJet had not properly supervised its maintenance contractor or the contractor's procedures; that SabreTech employees *failed* to properly prepare and package the volatile oxygen generators; and that the FAA *failed* to adequately regulate start-up airlines like ValuJet and *failed* to require smoke detection and suppression systems in DC-9 cargo compartments. [21] *The Washington Post* reported Kenneth Quinn, an attorney for SabreTech who ironically, was *formerly general counsel of the FAA*, as saying that "prosecutors were going down a dangerous path in applying criminal standards to an accident...Justice is not served by today's decision to attempt to criminalize human error."[22] Quinn was quoted in *The Miami Herald* as saying: "These were well intentioned individuals who made honest mistakes that actually pale in comparison with the failures of the FAA and ValuJet in this tragedy." As we shall see in subsequent sections of this chapter, it was not the first time that the notion of *regulatory and administrative failure* captured in the NTSB's report on ValuJet would be raised.

For an appreciation of the enormity of the task of regulatory oversight, a brief perspective of the U.S. air transportation industry is appropriate. The industry enplanes approximately 60 percent of the world's commercial air passengers and has the world's largest and most active population of General Aviation (GA) pilots. With regard to "personal" GA flights, according to the FAA's National Aviation Safety Data Center (NASDAC) in 1998, there were a total of 441 accident fatalities for the category of personal GA flights. For the first six months of 1999, there have been 164 such fatalities. Generally, except when accident fatalities occur among celebrities, such as the fatal flight of singer John Denver early in 1999, these accidents hardly make news. At the time of this writing, the NTSB began investigating the tragic accident on Friday night, July 16,1999, involving John F. Kennedy, Jr., his wife and sister-in law, while piloting his private plane. The Kennedy accident has resulted in intense media attention on pilot training regulations, the high accident fatalities among GA "personal" flights and the enormous economic and societal costs associated with the freedom to fly one's personal airplane. The NTSB is expected to issue a report on the cause (s) of the accident sometime between January – April 2000. There is strong suspicion that the report will address the issue of licensing standards, levels of training and academic knowledge of aviation psychology as conditions for private pilot licensing and bring about a re-examination of a regulatory oversight philosophy that, in the name of freedom, allows inexperienced, "low flying time" pilots to make *their own decisions* to fly under the twin challenges of nighttime flying over the ocean.*

Forecasts in 1994 are that the of U.S. commercial air passengers will double within 18 years to more than one billion passengers annually.[23] The airspace system *daily* handles more than 174,000 takeoffs and landings at airports across the nation, and *routinely* carries approximately 1.7 million passengers *safely* to their destinations.   Arnold Barnett, Professor of Operations Research, Sloan School of Management, Massachusetts Institute of Technology has quantified the safety of the system.   According to Professor Barnett, the death risk per US Domestic flight, 1987-1996 was 1 in 7 million. "At that level of risk, if a passenger chose one flight at random each day, the passenger would on average go for 21,000 years before succumbing to a fatal crash."[24] According to preliminary FAA figures, in 1998, major U.S. airlines flew without the death of a single passenger.   For the Air Transport Association's David Fuscus, while acknowledging that constant vigilance was necessary to keep fatalities down, particularly since air travel is up, this, for was a "testament to how safe commercial aviation is." For Mary Schiavo, the former DOT Inspector General who has been highly critical of the industry and its regulations, "this should not be seen as a cue for anybody to let up on safety." Flight Safety Foundation president, Stuart Matthews said accidents themselves were rare events and a single year was a poor way to analyze the data.[25]   A central thesis of this essay is that safety is more than the absence of accidents and it is a fatal flaw in communicating safety risk to the public to rely solely on sophisticated equations built on historical accident data. Essentially, what's being done is the making of a model about what's likely to happen in the future based on probabilistic studies about what happened

---

* Correspondence with an experienced pilot and engineer offered that when flying at night and you turn from a lighted seacoast to open ocean, it is like flying into a bottle of ink. You suddenly find yourself in instrument flying conditions. This is no problem if you are instrument rated and maintained proficiency. However, if you are not, vertigo strikes, the plane enters a turn and loses altitude. The inexperienced pilot's response is to pull back on the wheel to compensate for the loss in altitude. This tightens the turn and increases the rate of descent until the plane strikes the surface.

in the past.   When we examine the tenets of the "System Safety
Engineering School," the limitations of such an approach should become
more apparent.

Although the accident rates among air carriers is remarkably low, a
widely quoted Boeing study has concluded that if applied to the projected
future growth in air travel, the low accident rate will produce a loss
somewhere in the world every 8 days.   In the U.S., today's very low
accident rate would produce a fatal hull loss about every 3 months.   Add
non-fatal hull losses and cargo incidents, and "the U.S. could have
significant airline accidents in the news every 3 weeks, year after year.
Public demand for "action" would be so great that the industry likely could
not survive as we know it today."[26]   Similar warnings and concerns, albeit
in a slightly different context, over aviation safety were issued in 1970 in
John Godson's book, *Unsafe at Any Height,* twenty-five years earlier at the
beginning of the jumbo jet era.

The ominous warnings about future accidents with the approaching
era of jumbo jets that were voiced by John Godson were realized four years
later on March 3, 1974.   The crash of the McDonnell Douglas DC-10 some
25 miles north-north-east of the French capital became the center of
controversy following the Turkish Airlines Flight 981 near Ermenonville
that took the lives of 346 people including 12 crewmembers in what was
the world's largest civil airplane disaster.   The air transportation disaster
that had been feared since the start of the jumbo jet era – a non-survivable
crash involving a heavily-loaded wide-bodied aircraft - was used by
industry critics as an example of corporate ineptitude, design shortsideness
and government regulatory laxity.[27] The plane, Ship 29, a DC-10, with its
original faulty cargo doors, was the focus of the 1976 investigative report,
*Destination Disaster: From the Tri-Motor to the DC-10—The Risk of
Flying.* The book, written by a team of three veteran London *Sunday Times*
reporters (Paul Eddy, Elaine Potter and Bruce Page) provided the complete
history of the ill-fated plane and its record of corporate failure and Federal
regulatory irresponsibility.   It described in detail how in 1969 McDonnell-
Douglas failed to tell the FAA of potential malfunctions.   It also describes
the results of tests conducted in May 1972 in which the cargo door blew out
and the floor collapsed. Sadly, it recounts how after a near disaster over
Windsor, Ontario on June 12 1972 the General Dynamics engineers
predicted a fatal crash but were not permitted to act on their knowledge.   It
also describes a lax regulatory policy of reliance on the manufacturers to
police themselves and permitted Ship 29, to leave the factory unmodified,

despite controls and records indicating the contrary. That accident was preventable since the cause of the crash had been specifically predicted.[28]

The ensuing lawsuit, *Hope et al. v. McDonnell Douglas et al.,* U.S. District Court, Central District of California, yielded the largest compensation ever won in aircraft litigation. According to the team of *Sunday Times* reporters, the settlement agreement was *believed* to have resulted in: "the U.S. Government contributed a fixed sum believed to be $3 million; Turkish Airlines agreed to pay an average of $30,000 for every passenger killed, a total of $10,050,000: General Dynamics agreed to pay a fixed percentage—between 15 and 20 percent of the total compensation bill and McDonnell Douglas agreed to pay the balance." Lloyds of London carried the majority of the insurance risk for all three corporations. The authors also noted that in 1958, Judge Pierson Mitchell Hall began hearing his first major air-crash case. It took eight years to settle and the experience left him with a profound determination to "do a lot of innovating." By 1974 he had tried eight other major air-crash cases and settled each of them, on average, in two years. He did this by insisting on rigorous schedules for "discovery." This process very early determined that 346 people were killed because the rear cargo door was unsafe. Therefore, Turkish Airlines Ship 29 was undoubtedly a "defective product." The judge had also set, at an early stage in each case, a firm trial date. The authors also observed that Judge Hall's ruling on August 1, 1975, would likely become a landmark in air-crash litigation. Judge Hall has long held that air crashes cases should be tried under federal law to achieve a measure of consistency. The judge said that because the federal government wrote the laws, which govern aviation in the United States, it had an "interest" in all air-crash litigation. That interest was just as great as the interest of a state which, in wrongful-death law, was primarily concerned with deterring the manufacture and sale of defective products within its borders. Weeks before the ruling, the defendants had announced they had come to complete and confidential agreement among themselves as to sharing the damages of all the plaintiffs. Judge Hall's opinion, *In re Paris Air Crash of March 3, 1974,* on the 203 suits involving 337 decedents arising from that crash that were before his court became enshrined in aviation law. This was a remarkable departure from the liability limits contemplated by the 1944 Chicago Convention.

Destination Disaster contains extracts from a memorandum, written June 27,1972, by F.D. Applegate, Director of Product Engineering, Convair Division of General Dynamics Corporation (subcontractors to

McDonnell Douglas for detail design and construction of the DC-10 fuselage):[29]

> The potential for long-term Convair liability on the DC-10 has been causing me increasing concern. ...The airplane demonstrated an inherent susceptibility to catastrophic failure...in ground tests.... It seems to me inevitable that, in the twenty years ahead of us, DC-10 cargo doors will come open and cargo compartments will experience decompression...and... result in loss of the airplane. It is recommended...to persuade Douglas...to incorporate changes...which will correct the fundamental cabin floor catastrophic failure mode....

The authors of *Destination Disaster* then entered into a discussion of the legal and moral responsibility laid upon the officers of a corporation and the "plain human duty" to warn others of danger to life. Corporate officers have a duty to safeguard the corporate interest, and "if the duty to give warning can legitimately be placed elsewhere, they may be entitled, or expected, to remain mute.... that was the course that Convair chose to adopt." The book provided a comprehensive summary of the DC-10 accident. In 1968, while the DC-10 was still being designed, engineers from the Rijksluchtvaartdienst (RLD) the Dutch equivalent of the FAA issued repeated warnings about the integrity of the passenger compartment floors in jumbo jets. In 1969, a failure modes and effects analysis (FMEA) performed found nine potentially hazardous failure sequences. The exact problem that led eventually to the accidents was not identified by the FMEA, but the analysts did demonstrate the danger of the door design without a totally reliable fail-safe locking system. In a textbook case of the design/regulation nexus that underscored the lack of regulatory process integrity, the aircraft was "certified" by the FAA although the FMEAs submitted did not include any mention of the possibility of the hazard related to a malfunction of the lower cargo doors. The continuing Dutch RLD concern was explicitly stated at an ICAO meeting in Montreal. In the Windsor incident, catastrophe was avoided only because the pilot had trained himself to fly the plane using only the engines because he was concerned about a decompression-caused loss of the control cables. After his near catastrophe, the pilot recommended that every DC-10 pilot be trained for this eventuality. McDonnell Douglas never did this. Instead, they attributed the incident to human failure on the part of the baggage handler and not to any design error.[30]

Almost immediately following the American Airlines incident, the locking mechanism was suspect.   This prompted the FAA's Western Region Office to begin preparing an Airworthiness Directive (AD), which would have required, as an interim measure, the placement of a small viewing port to enable the observation of the position of the latches. But in the now infamous 'gentleman's agreement', between then FAA Administrator John Shaffer and Jackson McGowen, at the time president of the Douglas division of McDonnell Douglas, the AD was downgraded to three Service Bulletins. This in effect meant that the effort would not be urgently pursued.[31] The use of non-regulatory procedures and agreements proved ineffective. The persistence of an FAA employee who worked in the Western Region, eventually caused the aircraft manufacturer to divulge that there had been about *one hundred* airline reports of the door failing to close properly during the ten months of DC-10 service. In September 1972, the Dutch RLD again sent a delegation to Los Angeles, to meet with both the FAA and the manufacturer to discuss their concerns. McDonnell-Douglas took the position that the DC-10 floor met all of the FAA air worthiness directives. The RLD replied that the directives were inadequate. ...The RLD certified the DC-10, but placed on record their reluctance in doing so. Finally, by February 1973, the FAA had changed its mind and decided that something needed to be done about jumbo jet floors. McDonnell-Douglas insisted that the chances of a cargo door opening in flight were "extremely remote." [32] *Destination Disaster* quotes at length conclusions from a report of an FAA ad hoc committee to investigate the history of the DC-10 and the role that the agency had played in its development. These have been condensed accordingly:[33]

> It was therefore incumbent on McDonnell Douglas to show that loss of the cargo door... was "extremely remote" for compliance with Federal Airworthiness Regulation 25...in the light of the two accidents, the level of protection and reliability provided was insufficient to satisfy the requirements of FAR 25. ... The agency has been lax in taking appropriate Airworthiness Directive action.... Many complaints have been received from a number of foreign airworthiness authorities on (voluntary compliance programs). The FAA has ignored those complaints.

In the U.S. House of Representatives, a report by the Special Sub-Committee on Investigations of the Committee of Interstate and Foreign Commerce, January 1975 concluded, "through regulatory nonfeasance, thousands of lives were unjustifiably put at risk."[34] In addition to underscoring the crucial issue of regulatory process integrity, brought about by this disaster, the philosophical issue of "certification" versus "safety" became a critical one for the future of aviation safety regulation. In effect, the self-evident truth was that a manufacturer or operator could legally be in compliance with the governing regulations yet still be unsafe. It is an issue that would undergo a full airing some two decades later in 1996 with the *ValuJet* tragedy.

Ralph Nader reviewed *Destination Disaster* and summed up the book by warning that similar catastrophes stemming from corporate and regulatory failure could happen again. Nader's continuing concerns over transportation safety led to a publication, *Collision Course: The Truth About Airline Safety (1994)*. In the book, he has cited a number of safety-related issues that his Aviation Consumer Action Project has deemed necessary to enhance aviation safety. The Nader book includes a key section on the case of John King, whistle-blower and former mechanic for Eastern Airlines who was fired from his job after he tried to notify the FAA, via the anonymous Aviation Safety Hotline, that the airline was systematically falsifying its maintenance records. The kinds of problems that John King tried to bring to the FAA's attention subsequently led to the criminal indictment of Eastern Airlines. The indictment contended that the fraud began in 1985, two years before King was dismissed for telling the world that Eastern was engaged in the kind of maintenance fraud that King had described. In New York District Court, in 1989, Eastern Airlines pleaded guilty to six counts of the indictment and was fined $3.5 million.[35]

The U.S. Department of Justice filed a criminal lawsuit, *United States of America, v. Edward Upton et. al.,* arising out of charges of falsification of airplane maintenance records for Eastern Airlines, Inc. The thirteen defendants, former employees of Eastern Air Lines, were charged with obstructing the administration of the law by, among other things, testifying falsely before the FAA regarding their knowledge of and participating in a conspiracy to falsify Eastern's maintenance records.[36] The issue of the financial health of the airline industry in the era of deregulation and the impact on safety was a topic of serious concern.

In 1988, the book *Air Travel: How Safe Is It?*, by Captain Laurie Taylor, a senior pilot flying world-wide routes with British Airways and a former Chairman of the British Airline Pilots Association, examined the interrelated factors of regulation, economic strength and safety and concluded that it was impossible to achieve an acceptable level of air safety without taking all into account. The book devoted an entire chapter to the issue of Costs Versus Safety in the context of the survival of new entrants in the competitive environment. The author included lengthy quotations from ALPA on its concerns over safety in the era of airline deregulation:[37]

> ...To take comfort in the regulations and enforcement responsibility of governments is to avoid the safety question...The primary factor in airline safety is cost. Maintenance of high safety standards is extremely expensive, whether measured in terms of equipment, facilities, maintenance reliability or by operating procedures and conditions. This fact is often taken for granted because it is rarely visible to the consumer of airline services. ...airlines have invested millions of dollars in on- board safety systems, personnel training ...it is simply the price... to achieve the highest possible degree of safety in airline operations... most certificated airlines today operate according to standards which far exceed the minimum required by government. .... Extraordinary vigilance by government should be mandatory in a deregulated environment to guard against a relaxation of safety practices ... .

Support for ALPA's gloomy view on the safety consequences of deregulation also came from the Flight Safety Foundation. The book cited a noteworthy quotation by the President of that organization, Mr. J. Enders, at its October 1985 meeting: "The architects of deregulation failed to apply fault-tree analysis technique which the aircraft designer is expected to apply to an aeroplane...with the result being sharp economic competition which cannot enhance safety." When we investigate the tenets of the "Reliability Engineering School," and the issues of aviation safety and efficiency, the significance of that statement will be appreciated. In 1997, a 2nd edition provided an updated review of all aspects of international air safety.

The Airline Deregulation Act of 1978 transformed the commercial aviation industry. After deregulation, numerous questions lingered about its impacts and the adequacy of existing Federal safety policies and programs. In 1986, John J. Nance, a practicing attorney specializing in aerospace law and a professional pilot, published the book, *Blind Trust*. The book was a stinging indictment of airline deregulation and how it jeopardized airline safety. Nance documented a number of failings of the FAA's regulatory process related to major preventable accidents (Downeast, Air Florida, Air Illinois and PBA). For Nance, airline safety was in a state of a "national crisis" with a significant deterioration in the margin of safety—the potential for fatal airline accidents. Nance argued that "Basically Congress said that it would deregulate economic factors in the airline business, but would *not* deregulate safety. Yet by failing miserably to address the question of what deregulation would do to the system and by failing to beef up the FAA's surveillance and enforcement (programs) … in almost direct proportion to the degree of economic freedom, Congress virtually guaranteed that safety would be compromised, and that the FAA would lose control."[38] His book was a "call to arms for changing the system."

In 1988, the U.S. House of Representatives' Committee on Public Works and Transportation and the Subcommittee on Government Activities and Transportation of the Government Operations Committee, asked the Office of Technology Assessment (OTA) to determine how well existing safety policies, regulations, and technologies meet the government's responsibility for ensuring safety in commercial aviation. The OTA issued a report, *Safe Skies,*[39] that became the blueprint for major recommended FAA organizational, new programmatic emphases and the establishing the preeminence of the safety function in FAA's mandate.   The report noted that the dual mandate "… to promote safety of flight...in air commerce through standard setting..." and "...to encourage and foster the development of air commerce" created an inherent tension between the two vital FAA safety activities—inspections, and managing and operating the air traffic control system.   The OTA report identified two key areas for enhancing air safety:

- Safety management improvements, including streamlining FAA's internal organization, increasing inspector and operating work forces, raising levels of expertise, and establishing the primacy of FAA's safety responsibilities to ensure a more powerful *system safety* program; (Writer's emphasis) and
- System operating improvements, including expanding air traffic Control *capacity* and capability; enhancing human performance; and upgrading weather forecasting, detection, and dissemination and air/ground communications. (Writer's emphasis).

The 1989 publication, *Unfriendly Skies,* provided a practical account of on-the-job experiences and startling revelations about unsafe practices by a troubled deregulated airline pilot. "There have been pilots at some airlines who have been threatened with job dismissal. The casebooks are filled with literally hundreds of instances in which pilots have been excoriated for claiming that their airplanes lacked airworthiness."[40]

In keeping with the theme that *safety is everyone's business, Collision Course,* in addition to addressing a wide range of aviation issues, also discussed the issue of protection for whistle-blowers. In 1989, congress passed the Whistle-blower Protection Act—to protect Federal government employees from retaliation for reporting unsafe or improper activities such as fraud, waste and abuse. Nader made the case for extending similar protection into the private sector. He stated that "Airline whistle-blowers need to be assured of anonymity and that they can report perceived safety violations without fear of reprisal, harassment, or job loss. Whistle-blowers often face protracted legal battles that they must often wage at personal expense."[41] Nader stated that whistle-blowers should be given immunity from job actions, except if it can be shown that the employee lied or should have known the charges were false. He acknowledged the danger of disgruntled employees making false claims, but "for the sake of safety," some accommodation must be made to permit employees to whistle-blow—even when they make a good faith mistake—without fear of retribution. Aviation whistle-blowers are in a front line position to observe practices that an absentee FAA inspector might never see.*

---

*For an appreciation of the dynamic interplay of the role of the FAA Inspector, safety advocacy, organization culture and the politics of airline regulation in aviation safety, see the following articles: "Grounded by politics at the FAA. How a safety inspector lost her dream job." And "Safety Last, FAA Inspectors Complain. They accuse bosses of treating Alaska Airlines with kid gloves." *http //www.seattle-pi.com/pi/local.* Friday, March 5,1999. The articles detail how Ms. Mary Rose Diefenderfer, was removed from her job after investigating Alaska Airlines for violating pilot training or certification rules. There does not appear to be any publicly available information on the FAA's position.

See also: General Accounting Office Report: *Aviation Safety: FAA's New Inspection System Offers Promise, But Problems Need To Be Addressed.*

A discussion of the merits or effectiveness of the FAA's Aviation Safety Hotline, and by inference, the effectiveness of whistle blowing in furthering the cause of aviation safety is outside the scope of this book. However, passing reference is made to the irony of Office of the Secretary of Transportation support, and continuing congressional inaction on attempts "To amend title 49, United States Code, to provide protection for airline employees who provide certain safety information."[42] In a letter, to the Chairman of House Transportation Committee, the Secretary of Transportation stated that: "The enactment of H.R.3187 would provide an added inducement to share that type of information with the FAA." The effort failed to obtain legislative passage. For whatever reasons, neither the airline industry nor its congressional allies want whistle-blower protection. This remains an open debate with contradictory propositions of corporate authority and social responsibility.

At the time of this writing, Nader's efforts on behalf of aviation consumers are continuing. On July 3rd, 1998, his Aviation Consumer Action Project filed a lawsuit against the FAA for certifying the Boeing 777-300, the world's longest commercial airplane, with a capacity of 550 passengers, without holding a full-scale passenger evacuation test.[43]

In 1997, Mary Schiavo, the former Department of Transportation Inspector General, turned whistle-blower, published a best seller, *Flying Blind, Flying Safe.* It received widespread news media attention and stirred controversy about aviation safety. Her book was published shortly after the terrifying air tragedies in 1996. The book, in addition to exposing a regulatory mindset perceived to be at odds with the goals of airline passenger safety, called for the making of safety information available to the flying public.[44] To underscore the importance of safety information to aviation consumers, in 1996, Senators Ron Wyden and Wendell Ford wrote the FAA urging the agency take action to educate the public better about the safety of the aviation system and make important information about aviation safety more easily available to consumers. The agency's response on the issue of public information and education was positive and shortly thereafter initiated a quarterly compilation of all FAA enforcement actions against regulated aviation entities that involve safety and security issues. The agency also initiated a public education campaign to explore more effective ways of communicating with consumers about aviation safety.[45]

Report # RCED-99-183. July 7th 1999. Also, *USA Today,* "Report: Jet Safety Program Falls Short." July 7th 1999.

A persistent theme in the emergence of the "Transportation Tort Law School" of safety has been the role of single-minded individuals leading biblical David attacking the Goliath of transportation services corporations and a lethargic regulatory mindset, in the quest for personnel and personal safety. These battles which naturally began first against rail and then automobile sectors have resulted in the repudiation of the wretched doctrine of "buyer beware" *caveat emptor* in providing transportation services. The legislative initiatives have had a profound effect on new patterns in safety thinking and the emergence of government safety regulation. Since 1893 with the passage of the *Railroad Safety Appliance Act,* the eternal question remains the degree of regulation. Ralph Nader's thoughts and actions are truly representative of the "Transportation Tort Law School" of safety. He has been named one of the 100 most influential Americans of the 20th century by LIFE Magazine. He is the founder of Public Citizen, the nation's largest consumer advocacy organization.   Nader has proposed establishing a Tort Museum to honor Tort Law. The idea of the museum is to show how tort law developed and how it can be a major benefit to society, serving as a deterrent to unsafe products.   As he envisioned it, the 7,000-square-foot museum would contain a full-sized courtroom where visitors could participate in mock trials.   Specialty rooms, for topics such as failed medical devices that continue to occupy front-page media stories would be covered.   The general counsel to the Civil Justice Reform Group, which has championed product liability legislation for the past several years, noted that: " If anything or anyone has become famous because of American tort law, its Ralph Nader." [46]

Concluding Remarks

This Chapter has highlighted the century-old historical role of Lorenzo Coffin in transportation safety. It serves as a reminder of how far concern over safety has developed through tort law as well as underscoring the consequences of not adhering to "Best Practice." Following the historic GM Chevrolet Malibu Case, *The Times* noted that " (although) American damage suits are full of bad law, exaggerated liability, histrionics and random lottery numbers...such swingeing penalties do, however, have the merit of forcing big corporations to give safety top priority. Customers who have to trust big suppliers' technology need that protection. GM *et al* have had to learn since the days of Ralph Nader. (By contrast) In Britain...the lack of any accountability ...sends the opposite message. Too many companies still think they can afford to be complacent and put pressure on middle managers regardless."[47]

Here follows some practical admonitions stemming from society's heightened awareness of safety risk management, including the message being sent by jurors about product liability, with far reaching implications for aviation safety practitioners:

- **To Understand The Nature of Risk**, Charles Perrow, in his book, *Normal Accidents* (1984), made the following point: first differentiate between the *imposition* of risks by profit-making firms who could reduce the risk and the acceptance of risk by the public where private pleasures are involved. The poignancy of that statement was evident among the California jurors who made the largest award ever ($4.9 billion) involving a defective product. One juror was quoted as saying: "We're telling GM that when they know that something...is going to injure people, then it's more important that they pay the money to make the car safe than to come to court and have a trial."

- **Pencil Whippers Beware!** Aircraft Mechanics who may be pressured by the Vice- President for Maintenance or mid-level manager, to bypass prescribed work steps, and falsify documents indicating work had been completed (a practice known in the aviation industry as "pencil whipping") should be mindful of the historic aviation legal precedent of criminal charges and murder indictments in the SabreJet case. Those charges are meant to bring accountability for the loss of 110 lives on ValuJet Flight 292 in May 1996 in Florida. As a result of the federal charges, SabreTech faces about $6 million in fines and restitution, and the three employees face up to 55 years in prison and more than $2.7 million in fines and penalties.

- **Aviation Safety Hotline Callers Beware!** The FAA maintains a confidential reporting program for safer skies. This 24-hour Hotline (1-800-255-1111) makes safety everyone's business. Although callers are protected by the Privacy Act, however, aviation industry employees should be mindful that the U.S. government does not offer whistle-blower protection for industry employees who report violations of the Federal Aviation

Regulations as well as suspected unapproved parts, non-adherence to operational procedures and unsafe aviation practices.   Congress has repeatedly blocked legislative attempts to provide protection for airline industry employees who provide certain safety information.

Notes

[1] *The Guide to American Law,* (New York: West Publishing Company, 1984), Vol.10.

[2] This entire section has been paraphrased from Stewart H. Holbrook, *The Story of American Railroads* (New York: Crown Publishers, 1947), "The Airbrake Fanatic,"   pp. 289-300; Stewart H. Holbrook, *Lost Men of American History*   (New York: The Macmillan Company, 1947), "Discontent: The Mother of progress," PPS. 259-266; and *Who Was Who In America,* Vol.1, 1897-1942, (Chicago, IL: Marcus-Who's Who Inc., 1968), 7th Printing, p.238.

[3] Rayford W. Logan and Michael R. Winston, *Dictionary of American Negro Biography,* (New York: Norton & Company, 1982), p.453.

[4] U.S. Department of Transportation, "Transportation Secretary Slater Unveils National Education Initiative," DOT 81-97, May 30, 1997.

[5] Ralph Nader, *Unsafe At Any Speed*, (New York: Grossman Publishers, Inc., 1956), p. x.

[6] IBID., p. 9.

[7] IBID., p.34.

[8] IBID., pp. 343-346 *passim*

[9] IBID., LXXXV- LXXXVII passim.

[10] This entire paragraph is summarized from 119 Cal. App. 3rd 757: 1981 Cal.App. LEXIS 1859.

[11] Perrow, Op.Cit. p. 309.

[12] *Notre Dame Lawyer,* Notre Dame University: School of Law. 1979/06, pp. 911-924, "Corporate Homicide: A New Assault on Corporate Decision-Making." NHTSA Report, HS-028 688.

[13] *University of Detroit Journal of Urban Law,* Detroit University, School of Law, 1979. "Beyond Products Liability: The Legal, Social, And Ethical Problems Facing The Automobile Industry In Producing Safe Products." NHTSA Report, HS-030 349.

[14] *The Wall Street Journal,* "Sudden Acceleration Probe of Audis Finds No Defect," C8; 5, July 14, 1989.

[15] *The Wall Street Journal,* Law, B6; 5, December 12, 1989. See also *Wall Street Journal,* "Audi of America Agrees to Recall 5000-Model Cars," January 16, 1987.

[16] "Carjack Victim Succeeds in Fight For Safety Latches in Trunks," *The Washington Post,* June 19,1999, p. E1.

[17] "Proposal to cut hours truckers may work," *USA Today,* August 6-8, 1999, p. 1.

[18] "Youth Killed by Dump Truck Will have a Legacy of Reform," *The Washington Post,* August 4,1999, Metro, p.1.

The Coopers have proposed to the industry eight disarmingly simple steps that would incur no additional so-called "regulatory burden" - a favorite rallying cry of advocates of limited governmental intervention in the marketplace. These are inexpensive to administer and should give readers pause:

1. Require documentation of: all necessary licenses and permits; any accidents or citations involving trucks or on-duty drivers for the past year; the last required federal inspection; adequate liability insurance; current federal safety report; a pledge not to pay speeding tickets for drivers.

2. If any of the above information changes or expires during the project, the trucking firm must update its records.

3. Before work starts, the excavation company shall certify trucking company compliance.

4 Before any driver starts work on a project, the excavation company must inspect the driver's commercial license for currency and appropriateness for the job.

5. The excavation company must identify routes for drivers that minimize the use of streets through residential areas.

6. The excavation company must not require or recommend that drivers take any other routes

7. If the excavatioɪ; company uses how long a trip takes to measure a trucker's performance, the standard shall be the "reasonable" time it takes to drive the recommended route.

8. The excavation company must not load any truck with more weight than the maximum allowed in any jurisdiction in which the truck is driven.

[19] "GM Ordered to Pay $4.9Billion in Crash Verdict," *The Los Angeles Times*, Saturday, July 10,1999, Section: Part A. *See Also:* "A $4.9 Billion Message," *The Washington Post*, July 10, 1999.

[20] "Employees Polled," Federal Aviation Administration, *Intercom,* June 23,1998.

[21] "Murder Charged In Plane Crash," *The Miami Herald*, July 14,1999.

[22] "Murder Charged In ValuJet Crash," *The Washington Post*, July 14, 1999.

[23] *FAA Aviation Forecasts*, Fiscal Years 1995-2006.

[24] Briefing to the FAA, "Aviation Safety: The Recent Record,"   Arnold Barnett and Alex Wang, March 26,1998.

[25] *CNN*, "No U.S. Airline Deaths in '98," January 6,1999.

[26] FAA Administrator David Hinson's address before The Wings Club of New York, May 1995.

[27] *Aviation Disasters*, Op. Cit., pp. 124-126.

[28] Paul Eddy, Elaine Potter, and Bruce Page, *Destination Disaster*, (New York: Quadrangle/The New York Times Book Co., 1976).

[29] IBID., pp. 14 –15.

[30] Leveson, Op. Cit., pp. 563-565.

[31] *Aviation Disasters*, Op. Cit., p. 125.

[32] Leveson, PP.566.

[33] *Destination Disaster*, pp.160 - 161.

[34] IBID., PP161.

[35] Ralph Nader/Wesley J. Smith, *Collision Course: The Truth About Airline Safety,* (TAB Books, McGraw-Hill, Inc., 1994), pp. 322-325.

[36] Volume 856 *Federal Supplement*, PP.727; 1994 U.S. District. LEXIS 8873.

[37] Captain Laurie Taylor, *Air Travel· How Safe Is It?,* (Oxford: BSP Professional Books, 1988), pp. 144-145.

[38] John J. Nance, *Blind Trust,* (New York: William Morrow & Company, 1986), p. 378.

[39] U.S. Congress, Office of Technology Assessment, *Safe Skies for Tomorrow· Aviation Safety In A Competitive Environment,* OTA-SET-381 (Washington, DC: U.S.Government Printing Office, July 1988).

[40] Captain "X" and Reynolds Dodson, *Unfriendly Skies,* (New York: Doubleday, 1989), p. 176.

[41] Nader, *Collision Course*, p. 325.

[42] Transportation Secretary, Federico Pena's letter to the Honorable Bud Shuster, Chairman, Committee on Transportation and Infrastructure, July 9,1996.

[43] *The New York Times,* July 5, 1998, PP.A 12. See also: *The Washington Times,* July 4, 1998, p. A4.

[44] Mary Schiavo, *Flying Blind, Flying Safe,* (New York: Avon Books, 1997).

[45] Correspondence: Administrator Linda Daschle, to Senator Ron Wyden, January 28, 1997.

[46] *The Washington Post,* July 9, 1998, "Nader Envisions a Tort Museum, Least Corvair and company be Forgot," p. E.1.

[47] *The Times,* Business News Commentary, August 28,1999, p. 25.

3 Reliability Engineering School

"Whenever the reliability of a product affects human life the reliability problem usually becomes part of a large issue called safety."

Martin L. Shooman, Ph.D.

Hardware Reliability

For an understanding of the thinking that characterizes the "Reliability School" of safety, we begin our study by citing the publication of Martin L. Shooman's widely respected 1968 classic: *Probabilistic Reliability: An Engineering Approach.* Many engineers today credit Shooman of Brooklyn Polytechnic Institute, with publishing a truly comprehensive text in the then newly emerged field.[1] However, the roots of "Reliability" go further back to the early industrial engineering work of Frederick W. Taylor's: *The Principles of Scientific Management* (1911). He is often called the "Father of Scientific Management." Taylor's dominant concern was for efficiency and the most economical way of accomplishing routine work. "Taylorism" nurtured the growth of America's early industrial assembly line production methods that are rooted in reliability. The Shooman textbook included an excellent bibliography on reliability models and quality control. Since the mid-1950s much work has been done on reliability analysis and the titles Reliability Engineer and Reliability Group were born. Several texts have appeared on the subject of reliability, and college and industrial courses on reliability have been initiated. Publication in 1971 of Bertram Amstadter's *Reliability Mathematics* served the needs of the undergraduate or practitioner who has had only an introductory course in statistics/statistics quality control. In a single ready reference, this applied rather than theoretical text provides useful tools for addressing reliability statistics problems.

In the context of this chapter, it is important to introduce a book written in 1984 by Professor Henry Petroski of the School of Engineering at Duke University. The book, *To Engineer Is Human,* provides an introduction to technology and an understanding and appreciation of engineers and engineering without the benefit of an engineering or technical education. It examines the role of failure in successful design. The concept of failure—mechanical and structural failure—is central to understanding engineering, for engineering design has as its first and foremost objective the obviation of failure. Petroski has framed the concept and consequences of "failure" as follows:[2]

> Failures appear to be inevitable in the wake of prolonged success, which encourages lower margins of safety. Failures in turn lead to greater safety margins and, hence, new periods of success. To understand what engineering is and what engineers do is to understand how failures can happen and how they can contribute more than successes to advance technology.

The book contains an excellent discussion on the *factor of safety* concept in dealing with all the uncertainties of engineering design and construction. The discussion is supported by analyses of cases of catastrophic structural accidents involving suspension bridges. It also includes an analysis of the Kansas City Hyatt Regency Hotel walkway collapse. The catastrophic event of July 17,1981 "became a synonym for the greatest structural tragedy in U.S. history. According to Petroski, the factor of safety is a number that has often been referred to as a "factor of ignorance," because it provides a margin of error that allows for a considerable number of corollaries to Murphy's Law to compound without threatening the success of an engineering endeavor. The essential idea behind a factor of safety is that a means of failure must be made explicit, and the load to cause that failure must be calculable or determined by experiment. This would indicate that it is *failure* that the engineer is trying to avoid in the design, hence the interest in the reason why failures of real structures occur. *The factor of safety is calculated by dividing the load required to cause failure by the maximum load expected to act on a structure.* Factors of safety are intended to allow for a structure (such as a bridge) built of the weakest material to stand up under the heaviest imaginable traffic. The objective of the designer is to make a structure tough rather than fragile. Since excessive strength can be uneconomical, unnecessary and aesthetically unattractive, engineers must make decisions about how strong is strong enough by considering architectural, financial, political, and competitive market factors as well as structural ones.[3] For Petroski, the paradox of engineering design is that successful structural concepts devolve into failures, while the colossal failures contribute to the evolution of innovative and inspiring structures. However, when we understand the principal objective of the design process as obviating failure, the paradox is resolved. For a failed structure provides a counterexample to a hypothesis and shows us what cannot be done, while a structure that stands without incident often conceals whatever lessons or caveats it might hold for the next generation of engineers.[4]

The 1993 publication of Scott D. Sagan's, *The Limits of Safety,* examined the scholarly literature about complex organizations and identified two general competing "schools of thought" with differing theories on the causes of accidents. He identified a "high reliability theory," based on the work of a group of scholars at the University of California, Berkeley, High-Reliability Organizations project. The "high reliability" proponents argue optimistically that extremely safe operations are possible, even with extremely hazardous technologies, if the appropriate organizational design and management techniques are followed. The second school, "normal accidents theory," led by Perrow, presents a much more pessimistic prediction that serious accidents with complex high technology systems are inevitable.

The high reliability theorists are in agreement with the professional risk analysts and engineers who build nuclear power plants, commercial aircraft, oil tankers, petrochemical factories and other potentially dangerous high-technology systems that serious accidents with hazardous technologies can be prevented through intelligent organizational design and management. The high reliability theorists also cite the FAA's air traffic control system as an example of "design and management of hazardous organizations that achieve extremely high levels of reliable and safe operations." The common assumption of the high reliability theorists is that properly designed and managed organizations "can compensate for well-known human frailties and can therefore be significantly more rational and effective than can individuals." For Sagan, these organizational theorists fit into the tradition that W. Richard Scott has called the "closed rational systems" approach. High reliability hazardous organizations are "rational" insofar as they have highly formalized structures that are oriented to the goal of extremely reliable and safe operations. Sagan's work has summarized the *four major factors* that are seen by the high reliability school as contributing to high degrees of reliable operations and safety. These factors are: the prioritization of safety and reliability as a goal by political elites and the organization's leadership; high levels of redundancy in personnel and technical safety measures; the development of a "high reliability culture" in decentralized and continually practiced operations; and sophisticated forms of trial and error organizational learning.[5]   The book provides a convincing account of many political barriers to learning, in view of conflicts over narrow parochial "turf" interests lead to faulty reporting if incidents, secrecy and other aspects of administrative behavior.

In *Reliability And Risk Analysis*, (1993), M. Modarres provided an introduction and explanation of the practical methods used in reliability and risk studies, and discussion of their use and limitations.   For Modarres, "Reliability" has dual meanings: probabilistic and deterministic. In discussing the probabilistic nature of "Reliability," he offers the definition that it is *the ability of an item (product, system etc.) to operate under designated operating conditions for a designated period of time.* The ability of an item can be designated through a probability, or can be designated deterministically. Deterministic analysis deals with understanding how and why failure occurred and how it can be designed and tested to prevent its re-occurrence.   It includes an understanding of the physics of failure and the role of testing and redesign.[6]   Leveson offers the following definitions: *Reliability* is the probability that a piece of equipment or component will perform its intended function satisfactorily for a prescribed time and under stipulated environmental conditions.   *Failure* is the nonperformance or inability of the system or component to perform its intended function for a specified time under specified environmental conditions.[7]

Reliability engineering received increased emphasis with the rapid Cold War growth of military aircraft and missile systems development and became increasingly important to commercial-aircraft manufacturers. Reliability estimates (expressed as Mean Time To Failure, MTTF, and Mean Time Between Failure, MTBF) for the three basic subsystems of an aircraft: the structure, the propulsion, and the avionics have been satisfactorily developed.   These have been incorporated into the governing regulations *(Title 14, Code of Federal Regulations),* by the FAA and adapted by worldwide certification authorities and the International Civil Aviation Organization (ICAO).

Shooman recognized the fact that many differing approaches in reliability work varied with the individual and the organization:[8] For the mathematician faced with a reliability problem, it is an exercise in applied probability or statistics. For managers, the Reliability Group is generally a staff organization.   The quality-control experts view reliability as an extension of their efforts.   The components engineer tries to buy the best and most reliable parts, the system engineer looks for simplicity that will lead to a reliable design. For Shooman, no single approach is satisfactory, and the problem must be dealt with from birth-to-death, involving: raw materials and parts quality, conceptual design, detailed engineering design, production, test and quality control, product shipment, warehousing, operator skill and technique, maintenance, and product use. The importance of Shooman's book lies in the fact that he recognized the impossibility of treating such a broad spectrum of topics in one volume. He then developed in an integrated fashion several fundamental areas of interest to the engineer. The idea being that each system engineer need not be a reliability expert but must be aware of the available techniques of reliability engineering, reliability design, and reliability improvement. Reliability must be considered in evaluation of the system along with performance, weight, cost, volume, etc.

History of Reliability

Reliability was first recognized as a pressing need during World War 11. The preliminary steps taken were to establish Joint Army and Navy (JAN) parts standards and to set up the Vacuum Tube Development Committee (VTDC) in June 1943. At the close of the war, between 1945 and 1950, several studies revealed some startling results that served as an impetus for further investigations:

- A Navy study made during maneuvers showed that the electronic equipment was operative only 30 per cent of the time.
- An Army study revealed that between two-thirds and three-fourths of their equipment was out of commission or under repairs.
- An Air Force study conducted over a 5-year period disclosed that repair and maintenance costs were about 10 times the original cost.
- A study uncovered the fact that for every tube in use there were one on the shelf and seven in transit. Approximately one electronics technician was required for every 250 tubes. In 1937, a destroyer had 60 tubes; in 1952 the number had risen to 3,200.

One focal point of trouble was the vacuum tube. Following the VDTC, an airlines group set up a study in 1946 aimed at development of better electronic tubes. Parallel studies were conducted by Aeronautical Radio Inc. and Cornell University, in which thousands of defected tubes were examined. Between 1949 and 1953 Vitro Laboratories and Bell

Laboratories pursued similar studies on the failure of resistors, capacitors, transformers, relays, etc. In 1950 the Department of Defense established an ad hoc committee on reliability, which in1952 became a permanent group called the Advisory Group on the Reliability of Electronic Equipment (AGREE). An AGREE report was published in 1957, and followed by a specification on the reliability of military equipment.

According to Shooman, in performing the reliability analysis of a complex system, it is almost impossible to treat the system in its entirety. Hence, the logical approach is to decompose the system into functional entities composed of units, subsystems, or components. The subdivision generates a block-diagram description of system operation. Models are then formulated to fit this logical structure, and the calculus of probability is used to compute the system reliability in terms of the subdivision reliabilities.[9]

The Mastery of Risk

The public debate in recent decades over reports of health threats from the environment and reducing risks and controlling costs have been fed by the lack of universal agreement among scientists about which methods are best for assessing risk to humans. In this context, more than passing reference should be made to the successful 1979 activities of Lois Gibbs when the young 27 year-old self-described "housewife" challenged Hooker Chemical Co., New York State and 21,800 tons of chemical wastes dumped at Love Canal in 1979. Under her leadership, Love Canal homeowners mobilized and forced state and federal action to address the environmental disaster. In 1980, President Carter signed an emergency evacuation order that relocated about 7,000 families. Congress passed the Superfund Law to fund cleanup of sites around the country, including Love Canal, for which Lois Gibbs has been dubbed "Mother of the Superfund." In May 1998, Occidental Chemical Corp., Hooker's successor company, and the City of Niagara Falls settled two 19-year lawsuits. Occidental has paid more than $233 million in settlements to the state, the federal government and Love Canal ex-homeowners in recent years, without admitting any wrongdoing or negligence. Gibbs heads the Center for Health, Environment and Justice (CHEJ), a network of grass roots environmental groups, where people can get scientific information and training on fighting possible environmental hazards in their communities.[10]

In 1988, the book, *In Search of Safety,* by Dr. John D. Graham of Harvard University School of Public Health, examined issues of chemicals and cancer risk.[11] The Harvard Center for Risk Analysis has sponsored several workshops on risk analysis.

In 1999, one such workshop: "The Precautionary Principle: Refine It or Replace It?" was jointly sponsored by the Chemical Industry Manufacturers Association, and the Chlorine Chemistry Council, to stimulate scholars and practitioners to consider how the desire for precautionary action should be addressed. The *Precautionary Principle* has been widely used in European environmental regulation. Its fundamental premise is that prevention is better than cure. The 1992 Rio Declaration on Environment and Development stated: *"Where there are threats of serious or irreversible damage, lack of full scientific certainty shall not be used as a reason for postponing cost-effective measures to prevent environmental degradation."*

At first thought, certain sectors among aviation safety practitioners may question the applic ility of the precautionary principle to aviation safety. However, as we continue to trace the evolutionary spiral of safety thinking it should become apparent that a change has taken place from the historic regulatory mindset of: *Problem/Solution,* or a "fire-fighting" perspective to a "fire-prevention" or accident prevention approaches with broad implications for human and capital resource allocation.

In communicating risks to the public, it has become clear that algebraic equations or probabilistic valuations cannot answer the moral, ethical, philosophical, and socio-political questions posed by these problems.

In chapter 4, we will examine how the issue of risk assessment and risk management has been addressed by the National Academy of Sciences in a 1994 study, *Science And Judgment In Risk Assessment,* and in a 1996, the National Research Council report titled: *Understanding Risk.* In the latter report, risk assessment was defined as not a single, fixed method of analysis, but rather as:[12]

> A systematic approach to organizing and analyzing scientific knowledge and information for potentially hazardous activities or for substances that might pose risks under specified conditions.

Closely related to risk assessment is risk management, which the report defined as, the process by which the results of risk assessment are integrated with other information—such as political, social, economic, and engineering considerations—to arrive at decisions about the need and methods for risk reduction. According to the report, policy considerations derived largely from statutory requirements dictate the extent to which risk information is used in decision-making and the extent to which other factors—such as technical feasibility, cost, and offsetting benefits—play a role.

Risk is generally defined as a combination of the likelihood of an accident and the severity of the potential consequences. Early writers in the safety discipline such as H.W. Heinrich recognized the general lack of agreement on an acceptable definition of risk. There are two views of what constitutes "risk." For Heinrich, one is equated with uncertainty of loss, a subjective view; the other was a chance of loss, an objective mathematical view. Heinrich established a nexus between the techniques for risk handling and organizational type. However, regardless of organization type, the functions of risk were: (1) Risk identification; (2) Risk measurement and analysis; (3) Risk-handling techniques or methods; (4) Risk-handling technique selection; (5) Continual monitoring. These functions hold true for transportation safety practitioners, and the key to transportation safety risk management resides in Heinrich's timeless four basic methods available to a Risk Manager:

- Avoidance
- Reduction
- Retention
- Transfer

Risk management is the natural corollary of System Safety. Risk management is what permits us to manage hazards identified by the risk assessment process and thereby assure safety. The scrutiny of risk assessment methods and their expanding uses in the federal and regulatory process by interested parties in the regulated industries, environmental organizations, and academic institutions, has led to frequent and sharp criticisms of the methods used for assessing risk and of the ways in which the results have been used to guide decision-making. In his classic work, *Normal Accidents,* Perrow examined the "new science" of risk assessment because it counsels taking risks that are unacceptable and improperly evaluated for living with high-risk systems. For him, risk-benefit analysis, or risk assessment, while not as dangerous as the systems it analyzes, carries its own risks. In this very sophisticated field, mathematical models predominate: extensive research is conducted; and esoteric matters of Bayesian probabilities, ALARA principles (as low as reasonably achievable), and discounted future probabilities are debated in courtrooms as well as academic conferences. Some of the best scientific and social science minds are at work on the problem of how safe is enough? Perrow's critique of the field is that it acknowledges the difference between voluntary risks such as skiing and hang-gliding, and involuntary ones such as leaching of chemical wastes. It does not acknowledge the difference between the *imposition* of risks by profit-making firms who could reduce that risk and the *acceptance* of risk by the public where private pleasures are involved (skiing) or some control can be exercised (driving). "All are bundled up in a vague reference to market principles." Perrow cites the case of the Ford Motor Company decision not to buffer the fuel tank in the Pinto, and of the General Motors Company when it rejected warnings from engineers that the Corvair would flip over for the lack of a $15 stabilizing bar.[13] Chapter 2 of this t explored how these risk management decisions contributed to the "Transportation Tort Law School" and the exponential growth of government regulation of this sector.

The 1987 publication, *Managing Risk,* by Vernon L. Grose, provided managers with a "systematic loss prevention" methodology. The author developed a ranking technique, a "Hazard Totem Pole" –indicating a hierarchy of risks to assure that all risks are identified, evaluated, assessed and brought under control. However, the reality of our economic world of finite resources suggests that all risks should not be controlled since "it does not make sense to try to fix every situation where risks exists.[14] Grose's book provided insight into the old adage, "Safety is everybody's business," and why, unlike any other profession or trade, "safety" got that distinction. For Grose, that distinction came about because "no one really knows what to do about safety." So safety gets assigned to the four winds—in the hope that since everyone is responsible, no one can be blamed for accidents. This has been disastrous for classical risk management for three reasons: (1) Any job that can be done by everybody requires little if any talent; (2) Something that everyone is expected to do requires no budget or financial support; (3) Because of its apparently universal understanding, safety is not worthy of study or attention.[15] The reader is asked to consider Grose's assessment, in view of Hammer's prescient *Law of Safety Progress*, which states that:[16]

An unsafe product will bring on corrective action or drive its producer out of business, thereby raising the safety level of all such products.

As we examine the tenets of the "Reliability Engineering School," we understand why "Safety" may not have been deemed worthy of separate, specialized attention: The fundamental premise is that *safety is achieved through reliability; and reliability is further reinforced via redundancy. Hence the thinking that a reliable system is a safe system.* To underscore a troubling aspect of this type of thinking, in an aviation context, consider briefly that the authoritative text, *Aviation Disasters,* which provides the first detailed summary account of major commercial aviation disasters from 1950, lists no entries for Swiss Air.[17] The tragic *Swiss Air* 111 accident in September 1998 off the coast of Halifax, a "high reliability" airline with an impeccable safety record underscores the challenges of understanding "the dynamics of accident causation" outlined in the text, *Human Error,* by Dr. James Reason. The unique challenges in communicating aviation transportation risks to the public will be further explored in the discussion on aviation safety risk communication.

In the earlier cited 1993 text, *Reliability And Risk Analysis,* M. Modarres has laid out what he believes every engineer should know about the topic. Dr. Modarres states that risk analysis consists of answers to the following questions:[18]

- What can go wrong that could lead to an outcome of hazard exposure?
- How likely is this to happen?
- If it happens, what consequences are expected?

The 1995 textbook, *Risk-Based Management,* by Richard B. Jones, made a bold attempt at narrowing the gap between theory and actual plant operations and maintenance. In Dr. Jones' view, most maintenance and operations personnel were inclined to view the risk models that have been developed around rigorous mathematics, probability, and statistics as not being useful to meeting the challenge of guaranteed reliability at an affordable price. The book examined the ideas and methods that have been used to improve reliability measurement, and reduce risk in a wide variety of industrial situations. It discussed a wide variety of industrial situations and the major principles of reliability-centered maintenance - whose fundamental premise is to maintain system function by analyzing how a system functions and how it can actually fail.

For Peter Bernstein, the mastery of risk is the revolutionary idea that defines the boundary between modern times and the past: the notion that the future is more than a whim of the gods and that people are not passive before nature. Until human beings discovered a way across that boundary, the future was a mirror of the past or the murky domain of oracles and soothsayers who held a monopoly over knowledge of anticipated events. Risk management guides us over a vast range of decision-making, from allocating wealth to safeguarding public health.[19] Risk is not fate, it is choice. Bernstein's book, *Against the Gods,* is about those thinkers who revealed how to put the future at the service of the present. By showing how to understand risk, measure it, and weigh its consequences, they converted risk management into one of the prime catalysts that drives modern western society. For Bernstein, the ability to define what may happen in the future and choose among alternatives lies at the heart of contemporary societies.

It is beyond the scope of this text for a treatment of the complex topic of analytical tools. However, a brief description of analytical tools based largely on the earlier cited modern works of Bahr, *System Safety Engineering and Risk Assessment: A Practical Approach (1997)* and Stephenson's *System Safety 2000 (1991)* would be helpful. Beginning with Hammer's work in 1972, most texts in System Safety contain material on fault tree analysis (FTA). FTA is a major analytical tool of the safety discipline and both the Stephenson and Bahr texts are "practical guides" to acquaint practitioners with the strengths and weaknesses of FTA.

Safety Engineering Techniques

In 1981, Henley and Kumamoto's *Reliability Engineering and Risk Assessment,* provided principles of risk assessment, fault tree and event tree mathematics. In the early '90s, another text by Henley and Kumamoto, *Probabilistic Risk Assessment,* provided analytical methods for risk analysis applied to practical industrial problems. The text also included a detailed review of qualitative methods including failure modes and effects analysis, event trees, and cause-consequence analysis.

A number of transportation safety engineers/analysts believe that the lion's share of the safety evaluation process begins and ends with analysis of the functions (component pieces) associated with a product or service to be considered. In so doing, there is a great reliance on hardware reliability engineering tools such as Fault Tree Analysis (FTA), Functional Hazard Analysis (FHA), and Failure Modes and Effects Analysis (FMEA), buttressed by sophisticated quantitative analyses of equipment failure/reliability. The National Research Council's work of the Panel on Human Factors in Air Traffic Control Automation emphasized the point that designers strive to meet specific FAA procurement requirements (for e.g., build to 3 nines). Whereas this appears to be an admirable goal, absolute reliance on specified reliability should be treated with some caution for three reasons: [20]

- Like any estimate, a reliability number has both an expected value (mean) and an estimated variance. The variance is often ill defined and hard to estimate. When it is left unstated, it is tempting to read the offered reliability figure (e.g., $r = .999$) as a firm promise rather than the midpoint of a range.
- Objective data of past system performance reveal ample evidence of systems whose promised level of reliability greatly overestimated the actual reliabilities.
- Experience also reveals that it is next to impossible to forecast all inevitable circumstances that may lead a well-designed system to "fail," even given the near boundless creativity of the system engineer.

The panel believed that it is impossible to bring the reliability of any system up to infinity and therefore, one must introduce automation under the assumption that "somewhere, sometime the system may fail; system design must therefore accommodate the human response to system failure."

Fault Tree Analysis (FTA) is a graphical tool used in both the "Reliability Engineering" and "System Safety" schools. No discussion of FTA could be held without referencing the *Fault Tree Handbook*, published in 1981 by The U.S. Nuclear Regulatory Commission.[21] For Stephenson, "the primary advantages of FTA are that it does produce meaningful data to evaluate and improve the overall reliability of the system and that it evaluates the effectiveness of and need for redundancy." He concludes with a practical tip: use fault trees when you want to educate a nonengineer (particularly in a lawsuit) of how difficult it is for something to occur.

According to Nicholas Bahr, FTA is better known among reliability engineers. *FTA is a deductive approach* that, in addition to being a qualitative analytical tool, can also be quantified. The engineer postulates a top event-or fault-such as train derailment, then branches downward, listing the faults in the system that must occur for the topevent (accident) to take place. This disciplined and rigorous method forces the engineer to systematically list the various sequential and parallel events or fault combinations that, of necessity, must occur to experience the accident. Logic gates and standard Boolean algebra allow the engineer to quantify the fault tree with event probabilities and enable the probability of the accident or top event. What is important to remember is that FTA is not a model of all possible system failures or all possible causes but rather, a model of particular system failure modes and their constituent faults that lead to the accident.[22] In his tutorial on FTA, Bahr cites a few mistakes that are to be avoided in constructing, quantifying, and evaluating fault trees, among them are:

- Remember that the fault tree really looks at functions, not components.
- Remember that garbage in-garbage out. If the results of the quantified tree don't make sense, don't give them too much weight. It is much better to use quantitative trees for comparison, not as absolute number generators.
- Look closely at the failure modes to determine if they are independent or dependent. This is very important in probability manipulations.
- Be sure the top event is a high-priority concern.

At the 17th International System Safety Conference, the award winning paper: "Beauty And The Beast – Use And Abuse Of The Fault Tree As A Tool"[23] presented by R. Allen Long, Senior System Safety Engineer, described proper application and misapplications of the fault tree as a tool when evaluating complex systems. The author believes that to fully realize the FTA potential, the analyst must: 1) properly (and narrowly) define the Top Undesired Event; 2) arrange the tree into scenarios rather than "failures;" 3) use a consistent nomenclature to prevent confusing multiple failures as one failure or vice versa; and 4) use a computer program to perform cutset analysis. Above all, the analyst must be able and willing to work with a wide variety of engineering disciplines and subsystems. This includes the ability to see how the pieces fit in a system and to properly analyze the interactions at the interfaces.

Probabilistic Risk Assessment

The text, *Systematic Safety,* by Lloyd and Tye provides an understanding of the early underlying safety assessment concepts of aircraft systems in the U.S. and European regulations for transport aircraft. It points out that in the aviation industry, the advent of the "auto-land" system in the early 1960s precipitated the necessity to have some basic safety objectives, which could be applied to any system or function, and to develop particular detailed requirements. "It became apparent that by increasing the redundancy of the channels making up the system, increasingly high levels of safety could be secured, and the risk of accident was at least approximately predictable in numerical terms."[24]   James Reason's text, *Human Error,* provides an understanding of how those principles, pioneered by the aviation industry, have been successfully adopted by the nuclear industry. Reason's summary of the underlying philosophy of probabilistic risk assessment (PRA), its analytical techniques and application to the nuclear industry, is condensed in the following paragraphs.[25] In the U.K., the nuclear power industry, faced with the requirement to obtain public acceptance and an operating license, must demonstrate that their reactor designs will satisfy certain safety targets. These safety targets have been expressed as order-of-magnitude probabilities: less than 1 in 10,000 per reactor year for a limited emission.

PRA has two basic aims. (1) To identify potential areas of significant risk and indicate how improvements can be made; (2) To quantify the overall risk from a potential hazardous plant. At the heart of PRA are logical "tree" models of the plant and its functions. These trees take two basic forms: (a) fault trees that address the question: How can a given plant failure occur (e.g., a serious release of radioactive material)? And (b) event trees that answer the question: what would happen if a given fault or event occurs (e.g., a steam generator tube rupture or small loss of coolant accident)? In the case of a fault tree the starting point is usually a gross system failure (the top event) and the causes are then traced back through a series of logical AND/OR gates to the possible initiating faults. An event tree begins with an initiating fault or event and works forward in time considering the probabilities of failure of each of the safety systems that stand between the initial malfunction and some unacceptable outcome.

Reason further notes that the general structure of PRA was established in 1975 with the publication of the U.S. Reactor Safety Study, a weighty document, WASH-1400: *An Assessment of Accident Risks in U.S. Commercial Nuclear Power Plants.* Although closely tied to the nuclear power industry, PRA can be applied across a wide range of high-risk industries. Essentially, PRA involves the following process:

Identify the sources of potential hazards. (For a nuclear power plant, the major hazard is the accidental release of radioactivity material).

- Identify the initiating events that could lead to this hazard.
- Through the use of fault trees, establish the possible sequences that could follow from various initiating events.
- Quantify each event sequence. This process involves data or engineering judgement about two things: (a) the frequency of the initiating event, and (b) the probability of failure.
- Determine the overall plant risk. Naturally, this would be a function of the frequency of all possible accident sequences and their consequences.

Reason has pointed out that by focussing only on hardware failures, PRA has been criticized on a number of grounds. The logic of event trees demands that only conditional probabilities should be used, thus allowing for the preceding components of an accident sequence. PRA assumes independence of events and thus, "this conditionality is rarely recognized." PRAs have neglected the possibility of "common-mode failures, something that is considerably enhanced by the presence of human beings at various stages in the design, installation, management, maintenance and operation of the system."

Reason believes that despite this shortcoming, the development of a standardized PRA represented a major advance in reliability engineering. The underlying logic of PRA provides designers with an ability to identify and install redundant safety systems to prevent accidents. However, in 1979, after the Three Mile Island accident, the major failing of PRA was revealed in its inability to accommodate adequately the human failure and mistakes that led to the accident. This problem has been the stimulus for numerous attempts and techniques to numerically convert human error rates and thus evolved a new science of Human Reliability Analysis (HRA). In summary, high-risk technologists have attempted to find an HRA method that is comparable in terms of its predictive accuracy to what had been done for mechanical systems.

Human reliability is the probability that a person will correctly perform a system-required activity in a required time period and will perform no extraneous activity that will affect the correct performance. The PSAM (Probabilistic Safety Assessment and Management) conferences provide many excellent papers. HRA is the systematic analysis of that probability. There are several phases involved, including: (1) Task analysis; (2) Identification of the potential human, both cognitive and psychomotor, functions performed by the worker, (3) Quantification of errors. HRA is used as a tool to aid in the design of processes when tasks are human-action intensive or the human error is the primary initiator for the accident sequence. Although it is beyond the scope of this text to review all the HRA techniques, passing reference is made to one of the widely known means of providing human reliability data for PRA studies - Technique for Human Error Rate Prediction (THERP). Reason's text provides a summary of THERP as described in *Handbook of Human Reliability Analysis with Emphasis on Nuclear Power Plant Applications.* For Reason, the basic assumption of THERP is that:

> The operator's actions can be regarded in the same light as the success or failure of a given pump or valve. ...the reliability of the operator can be assessed ...the same way as an equipment item. ...Activities are broken down into task elements and substituted for equipment outputs along conventional reliability assessment.

Reason believes that it is important to see these HRA models in the context of their development. They came about to meet the demands of PRA analysts to understand the neglected human error contribution to system accidents. His book is designed to assist, for without an "adequate and conceptually meaningful error classification and a workable theoretical infrastructure, there can be little or no principled basis to the business of human reliability quantification." Increasingly, as more and more human factors specialists make their presence felt in this area the hope is for the emergence of more effective HRA techniques. Within the HRA movement, an effort has been underway since 1995 to develop a HRA knowledge-based expert system. HRA identifies where human errors are most likely, estimates the error rate for individual tasks, and highlights the most beneficial areas for system improvements. Common HRA techniques and associated databases have been collected and translated into an electronic format. Next, the procedural rules and data were extracted from those techniques, modeled using artificial intelligence, and compiled into individual modules. Finally, these modules were combined into an HRA-based expert system that will provide probabilistic estimates for potential human errors within various risk and hazard assessments. Essentially, the goal is to incorporate the best features of these techniques and incorporate them into an "Ideal type" model that would encompass issues of conceptual and empirical validity. [26]

The work of Elisabeth Pate-Cornell, Department of Industrial Engineering and Engineering Management, Stanford University, in the PRA area has been critical of the traditional approach to including human errors in PRA. The approach has been to simply assume *a priori* that they account for an arbitrary proportion of the failure risk (e.g., 80%) and to further assess the effects of these errors on the probabilities and the consequences of the various failure modes. Professor Pate-Cornell has analyzed the grounding of ships due to loss of propulsion and illustrated how the "80% rule" is misleading and can yield wrong priorities among safety measures.[27]

Redundancy and Safety

Sagan's work, *The Limits of Safety,* examined the numerous psychological studies that have demonstrated what we have all know from daily human existence: human beings are not perfectly rational machines, but rather operate with limited and fallible cognitive capabilities. Organizational theorists have long been aware of the limits of human rationality and in seeking to solve a basic puzzle, John von Neumann has asked the fundamental question: Is it possible to build "reliable systems from unreliable parts?" According to the high reliability organization theorists, the answer is a resounding yes and the key design feature of such organizations is redundancy. Sagan cites the work of the Berkeley scholars who have found redundancy to be critical to the success of virtually all the successful high-reliability organizations they have studied. One such effort has observed that "U.S. air traffic controllers continued to use a voice radio system as a redundant backup device to map the locations of aircraft, even after more accurate radar technology became available." He notes that Aaron Wildavsky's, *Searching For Safety,* is an exception to the embrace of redundancy by the high reliability theorists. Wildavsky's study of nuclear power plants argues that the addition of some redundant safety devices may add to safety, but that too many can decrease safety since "they get in each other's way." Sagan has summed up the high reliability organization theorists' view of redundancy as absolutely essential if one is to produce safety and reliability inside complex and flawed organizations.[28]

Leveson's states the problem of redundancy somewhat differently: "Redundancy may increase complexity to the point where the redundancy itself contributes to accidents."[29] The 1997 text by Duke University Professor, Kishor S. Trivedi, *Probability and Statistics with Reliability, Queuing, and Computer Science Applications,* agreed that one way to increase the reliability of a system is to use redundancy. "However, we should be aware of a law of diminishing returns: The rate of increase in reliability with each additional component decreases rapidly as (the number of components) increases."[30]

The "normal accidents theory," led by Professor Charles Perrow, presents a much more pessimistic conclusion about the risks of using hazardous technologies. This group has examined the same industries and has concluded that despite hard work and efforts to maintain safety and reliability, serious accidents are a "normal" result. From Sagan's perspective, although serious accidents in such organizations may be rare, "they are inevitable over time. The belief that intelligent design and management will result in complex organizations that are capable of safely operating hazardous technology is an illusion."[31] The normal accidents theory is both structural and political. Perrow has identified two specific structural characteristics of many organizations operating dangerous technologies—"interactive complexity" and "tight-coupling"—which make them highly accident prone despite the intentions of their leaders. The political dimension of the theory focuses attention on the interaction of endogenous and exogenous conflicting organizational interests. These conflicts can influence the frequency of catastrophic accidents or their interpretation and blame for failures. For Perrow, interactive complexity is a measure of the way in which organizational parts are connected and interact. "Complex interactions are those of unfamiliar sequences, unplanned and unexpected sequences, and either visible or not immediately comprehensible."[32] From the normal accidents perspective, each of the *four factors* that have been identified as contributing to high reliability are seen as ineffective, unlikely to be implemented, or even counterproductive. Often, competing organizational and individual objectives will remain and thus severely impair efforts to improve safety. "Adding redundancy does not necessarily enhance reliability because it also increases interactive complexity, encourages operators to run more risks, and makes the overall system more opaque."[33]

The competing safety perspectives between Sagan's postulated "High Reliability Theory" and "Normal Accidents Theory" are listed below.[34] The summary material provides a framework for thinking about safety and aviation system efficiency.

High Reliability Theory

- Accidents can be prevented through good organizational design and management.
- Safety is the priority organizational objective.
- Redundancy enhances safety: Duplication and overlap can make "a reliable system out of unreliable parts."
- Decentralized decision-making is needed to permit prompt and flexible field-level responses to surprises.
- A "culture of reliability" will enhance safety by encouraging uniform and appropriate responses by field-level operators.
- Continuous operations, training, and simulations can create and maintain high reliability operations.
- Trial and error learning from accidents can be effective, and can be supplemented by anticipation and simulations.

Normal Accidents Theory

- Accidents are inevitable in complex and tightly coupled systems.
- Safety is one of a number of competing objectives.
- Redundancy often causes accidents: it increases interactive complexity and opaqueness and encourages risk- taking.
- Organizational contradiction: decentralization is needed for complexity, but centralization is needed for tightly coupled systems.
- A military model of intense discipline, socialization, and isolation is incompatible with democratic values.
- Organizations cannot train for unimagined, highly dangerous, or politically unpalatable operations.
- Denial of responsibility, faulty reporting, and reconstruction of history cripples learning efforts.

Aviation Safety and Efficiency

The following bizarre newspaper reports of a near aviation accident underscores a major theme of this essay: *Safety is more than the absence of accidents*. The material is presented so the reader can place in context the preceding discussions on system reliability, redundancy, safety and organizational culture as well as to underscore the strategic economic importance of aviation safety.   The purpose is also to cite an example of how some incidents with no injuries or fatalities are potentially more likely than accidents to deliver worthwhile safety and efficiency outcomes.

On June 28 and 29, 1998 various newspapers reported an incident,[35] of a near-collision between arriving and departing planes at LaGuardia Airport in New York, that had occurred *two months* earlier on April 3, 1998. The bizarre incident occurred when an arriving US Airways plane came within a reported 20 feet of a departing Air Canada jet, some 200 feet above the runway intersection.

According to those press reports, an official of the air controllers' union said the incident began when a supervisor spilled a cup of coffee, distracting a controller who should have redirected the planes sooner. Instead, the controller turned from the screen to help clean up the spill.

The top FAA air traffic manager for the New York area expressed the concern that *"Our system is designed to prevent aircraft from getting that close. But in this case, it is clear the system did not work the way it does day in day out."* Equally telling was the view by the leader of the controllers union at LaGuardia that: "This happened because the FAA wants controllers to run planes as tightly as possible." A member of the Air Line Pilots Association (ALPA) Safety Committee said that this was "an exceedingly dangerous event, as close as you can get and not have an accident...it was dangerously unacceptable." The NTSB is investigating the incident.

In view of the fact that some of the statements appear to impact issues of safety, efficiency and air traffic density, found in the *FAA 1998 Strategic Plan,* it would be appropriate to examine those issues.

Judging from news media reports, the first eight months of 1999 have proved to be exceptionally busy for FAA investigation teams on high profile near mid air collisions. FAA rules define a "near-miss" as 500ft or less. On May 25, CNN reported a close call between a British Airways Concorde and an American Airlines 767 at John F. Kennedy International Airport. According to the FAA, the two planes came within a mile of each other in the air, violating a three-mile separation rule. The Concorde was one mile from the runway when the pilot decided to divert the landing due to poor visibility. The Concorde reportedly then veered into the airspace of a parallel runway where the American plane was taking off. A near-collision report filed by the American 767 pilots said the distance of the Concorde was closer to 500 feet vertically and 1500 feet horizontally.[36] On July 9, ABC News reported a bizarre incident that, if validated, served to underscore the fragility of an aspect (no single point of failure) of the air transportation system: "Construction Worker Zaps Miami Air Control Center."[37] A construction worker working on a drywall project in a communication and computer equipment room at the Miami Air Traffic Center "inadvertently" tripped an electric switch that shut down power to communications in the control center. That action caused controllers to lose radio contact with about 100 planes for 10 minutes when the backup radio system failed. On August 26, the FAA began an investigation of an incident involving two Concorde flights at John F. Kennedy International Airport. The planes, travelling at 225-250 mph, passed within 800ft vertically and about three-quarters of a mile laterally as one took off and one landed. FAA rules required the planes to be 1,000ft apart vertically and three miles laterally and characterized the incident as a Controller error.[38] BA's chief Concorde pilot said the incident was a "non-event."[39] An interesting aspect of these reported news events is the differing perspectives and definitions of what characterizes a "close call."

We have seen throughout this essay, that "Safety" embraces a number of issues ranging from cultural, management, engineering and economic. In this section, we shall concern ourselves with the latter. To assist in the discussion, and to provide an additional context for the forecasted increase in air travel and the FAA's strategic goals of achieving safety and efficiency, a Model of The Aviation System Capacity Problem, was constructed.[40] As can be seen, the mixture of projected increased air traffic growth, the search for increased air traffic management efficiency and the growing awareness of the strategic importance of airline safety, breeds natural tensions between market-determined principles and social/political principles of safety.

To provide the reader with a *snapshot* of the complexity of the management of the National Airspace System, a conceptual model (Figure 3:1) that has decomposed the capacity issue was developed. The decomposition of issues surrounding aviation system capacity and efficiency should allow policymakers to identify key variables to produce information that will help in understanding the problems of air traffic congestion. This is consistent with the airlines' objective to provide safe, low-cost air transportation to the maximum number of air travelers and the mission of the FAA to provide for a safe, secure, and efficient global system. The model is based on the economist's supply and demand perspective. As such, it examines the interdependent nature of several variables and assures that the social costs of safety and environment are addressed.

As can be seen from *Figure 3:1* on the following page, the desire for easy access to air travel that is both safe, reliable and timely, is a multifaceted issue that encompasses several factors including the building of new airports or expanding existing capacity. Airlines' operating cost factors as well as the social costs of aviation (noise, pollution and congestion) are naturally entwined.

The public debate about congestion, crowed skies and air safety has been very intense. As early as 1989, a *New York Times* editorial suggested that "spreading flights evenly throughout the day by charging higher takeoff and landing fees to discourage flights during peak travel periods" can relieve congestion.[41] A full decade later, *The Wall Street Journal* ran a front-page story on air travel congestion. The article cited a number of examples ranging from technology to weather and air traffic control and management practices that are contributors to the congestion problem, including the following: "with so many planes, delays are inevitable at some airports even in good weather. Newark Airport in New Jersey can handle a maximum of 48 arrivals an hour. But some hours, airlines schedule 55 to 60 arrivals, guaranteeing some late arrivals.[42] In an article featured in the trade publication, *Aviation Daily,* Carol B. Hallett, president and chief executive officer of the Air Transport Association, in response to criticism by an aviation consultant that blamed the airlines for congestion, pointed out:[43]

> Airlines design their schedules to meet customer demand, providing customers with service timed to respond to their needs that is what they are in business to do.

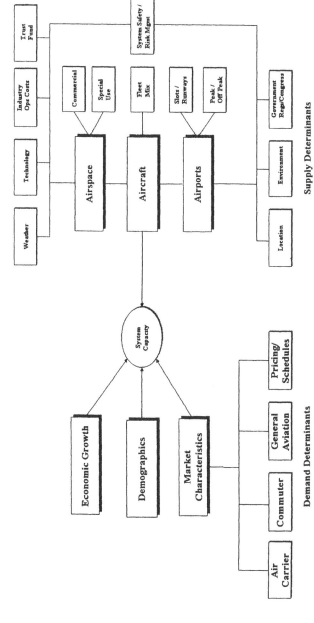

A Model of the Aviation System Capacity Problem

Figure 3:1 Capacity Model

The intensity of the public debate was further ratcheted when *Aviation Week & Space Technology* undertook a comprehensive examination of the "impending crisis." That article pointed out that each of the major stakeholders—the governments, the controllers and the airlines—have had a field day pointing out each others' shortcomings while minimizing their own contributions to growing problems."[44] It also developed an "Action Plan for Air Travel that proposed a number of steps to reduce congestion including organizational and federal budgetary reform, increasing the supply of airports and privatization of the FAA's air traffic services role.

A few days earlier, *The New York Times* had featured an article on Canada's private air traffic control system and noted that NAVCANADA "sought to run the system for the benefit of the users, instead of the convenience of the Government."[45]

Let us briefly examine the capacity problem. Often, the only way to achieve more capacity during peak travel periods is to increase the efficiency of existing runways. In November 1998, two weeks before the start of the busy travel period, *USA Today* featured a front page story on the FAA's policy of the simultaneous use of intersecting runways as a way of boosting traffic at the nation's clogged airports.[46] Since 1968, early trials with air traffic control procedures have permitted controllers to shuffle departing jets between arrivals on the same runway. Controllers bring arriving jets close together on parallel runways. One of the most successful ways to boost traffic has been operating two intersecting runways at full capacity at the same time under certain restricted operating conditions. This practice is permitted at 850 runways in 220 airports around the country. Runways typically have far more length than needed. If landings are made properly, a jet can stop well before the runway intersection.

Citing the 30-year *accident-free experience* with this operation, Federal and Air Transport Association officials call this a safe procedure. "No" says Capt. Randolph Babbitt, President of the Air Line Pilots Association, "they are just trying to squeeze out more capacity from the system." Capt. J.A. Passmore, British Airways chief of safety, in a letter to the FAA after an incident in May 1997 at Chicago O'Hare International Airport: "It is our view that this practice is inherently flawed." In that incident, a British Airways 747, loaded with 335 passengers and gathering speed for takeoff, was ordered to stop barely in time to avoid colliding with a smaller United Airlines 737 with 80 people aboard that was headed for a landing that pointed directly toward the jumbo jet's path. The United pilot had been cleared to land after promising to stop before reaching the 747's runway. However, strong winds prevented the United jet from stopping in time. The jumbo jet blew six tires, twelve brakes were locked and passengers shocked. Some pilots say that landings requiring them to stop before the end of the runway reduces that margin of safety: "What if weather or unforeseen circumstances force them to land too far down the runway to stop on time? What if an inexperienced pilot fails to understand the landing procedure and crosses a runway being used by a fully loaded jumbo jet? Or what if a pilot encounters trouble on the runway and is forced to gun the engines and lift off again? The jet could accelerate into an intersecting runway where another plane is taking off. These questions are classic System Safety accident scenario constructions. The FAA has initiated a number of analyses of the safety of these operations. Based on the results of these studies, the agency has made changes to the procedure to improve safety. What is clear from the public discourse is the existence of a risk communication vacuum.

Aviation Safety Risk Communication

The 1992 publication, *Risk: Analysis, Perception and Management,* by The Royal Society, reported that "the study and practice of risk communication is a relatively new development, with most of the relevant literature appearing after publication of the original Royal Society report in 1983. ... It is clear that several substantial issues, relating basic risk-perception research to public policy and decisionmaking, as well as to the process of risk management, are raised when risk communication is considered."[47] In that publication, Pidgeon et al offered that "risk perception involves people's beliefs, attitudes, judgements and feelings, as well as the wider social or cultural values and dispositions that people adopt, towards hazards and their benefits."[48]

The 1997 textbook, *Mad Cows And Mother's Milk,* written by Douglas Powell and William Leiss of McGill University, has provided seven case studies on risk communication failures between scientific experts and the public.[49] There are instructive lessons in communicating risk to the public. The authors have defined risk communication as "the process of exchanges about how best to assess and manage risks among academics, regulatory practitioners, interest groups, and the general public. Risk is the probability of harm in any given situation, and this probability is determined by two factors: (a) the nature of a hazard and (b) the extent of anyone's exposure to that hazard."[50] The 1998 proceedings of the International Conference on Probabilistic Safety Assessment and Management (PSAM 4) contained several papers on the topic of communicating risk to the public. Several conference papers suggested that often when risk assessments were presented they tended to reinforce communities' fears and concerns rather than allay them.

At the 1998 Transportation Hearings before a subcommittee of the Committee on Appropriations, House of Representatives, Chairman Frank R. Wolf expressed concern over the issue of safety metrics and in communicating aviation safety risk with the traveling public.[51] The Committee expressed concern that aviation safety statistics are produced in a "haphazard" way which responds to a particular incident or media exposure. Congressman Wolfe, citing the higher accident rate in Eastern Europe, Russia, Latin America and Africa, asked the FAA Administrator, Jane Garvey, if it would be a "good idea" to publish the safety records of foreign air carriers* so that individuals who are going to those countries would know the potential risk. The Congressman recalled the 1988 Pan Am, Flight 103 bombing over Lockerbie, Scotland, in which information on a bomb threat was not made available to the traveling public. Congressman Wolfe thought that safety and security information should be published on the Internet, that everyone could access. The issue of the travelling public's right to know the safety record of a particular airline (previously voiced by Ralph Nader and Mary Schiavo) was now ratcheted up. Administrator Garvey, cited the Boeing study on the occurrence of future accidents, agreed that the safety standards of the international community "is a very serious issue" although it is clearly less of an issue in the U.S. The Administrator acknowledged the complexity of the issue of airline rating and what kind of information is placed on the Internet but agreed that "the more information that we can give consumers, the better

---

* On August 5, 1999, The Department of Defense (DoD) and the Air Transport Association (ATA) signed an agreement that commits U.S. carriers represented by ATA who help fly American troops overseas, to assess the safety of all existing and proposed foreign code-share carriers. The agreement grew out of a 1985 charter crash in Newfoundland that killed 248 U.S. soldiers. The DoD leveraged its $1.2 billion travel budget to make U.S. airlines ensure the safety of their foreign partners who might carry U.S. troops. The agreement should help prevent commercial passengers from flying on foreign carriers with substandard safety programs.

off we are and the better off the consumer is." The Committee inserted an additional $1,000,000 into the FAA's budget to coordinate activities which will result in more effective and useful measurement of aviation safety nationwide.

Clearly, in recalling the Pan Am "situation," Congressman Wolfe has highlighted the issue of aviation consumer information. It may be helpful to briefly examine the communication aspect of that watershed event. According to the account in David Gero's book, *Aviation Disasters*, 16 days before the departure of Pan Am Flight 103 on 21 December 1988, an anonymous telephone message had been received at the U.S. Embassy in Helsinki. It warned that a sabotage attempt would be made within the succeeding two weeks against an American aircraft flying between Frankfurt and the United States. Word of the threat was passed on from the FAA to various American embassies, "presumably to give government employees a chance to make alternative travel arrangements if they chose, and to the carrier itself. But the general public was not made aware of the warning, on the rationale that such action would only serve to give the perpetrators publicity, and thus credibility, while potentially bringing financial harm to the airline industry. Besides, the threat had been dismissed as a hoax by some authorities."[52] As a result, all 259 persons aboard the aircraft and 11 on the ground perished. Five others were injured and more than 20 houses destroyed outright or damaged beyond repair. Gero's account reported that a week after the disaster, the British Air Accidents Investigation Branch confirmed what many had already suspected, that Flight 103 had been sabotaged. Pan American received harsh criticism for lax security in the wake of the tragedy, and both the airline and the U.S. government *for not announcing the prior warning.* Slightly more than a year later a similar threat made against a Northwest Airlines transatlantic flight was made known to ticket-holders. Although the aircraft reached its intended destination without incident, this marked a change in U.S. policy with regard to potential acts of terrorism.

The terrifying air accident of *TWA Flight 800* in August 1996 further underscored the need to draw a distinction between regulatory compliance for "certification" and "safety" when communicating risks to the public. It also crystallized the public debate on the pivotal accident prevention role of operational information sharing among regulators, accident investigators and the regulated in the overall safety equation. In the 1999 text, *Beyond Aviation Human Factors,* by Daniel E. Maurino, James Reason, Neil Johnson, and Rob B. Lee, the authors advanced an information-based "organizational model" for accident prevention. This is an alternative method to the traditional reliance on the actions or inaction of front-line operational personnel to pursue safety and accident prevention strategies in modern aviation. [*]

Consider briefly the following public reactions: The NTSB expressed "dismay and displeasure" when it learned that Boeing Co. had studied fuel tank problems some 16 years before similar problems in a similar jet apparently contributed to the explosion of *TWA Flight 800* – but failed to give the 4 volume report to safety investigators. For Senator Charles E. Grassley, Chairman of a subcommittee that oversees airline disaster investigations: "Boeing was under a legal as well as moral obligation to produce the study." For an attorney on the plaintiff's committee representing the families of *TWA 800* victims, "If Boeing knew about the problem, it should have done something about it."[53]

In May 1998, in the apparent absence of regulatory information on fuel tank problems, the FAA, in a watershed regulatory event, grounded[54] the entire fleet of Boeing 737s, the agency, in effect challenged the longstanding belief and traditional regulatory approach to "safety." Up to that time, "safety" had always been defined as being in compliance with safety airworthiness regulations. That historic action told us that government regulators do possess strategies that can be readily implemented. While this is hardly a startling conclusion, it is an instructive one. In so doing, the FAA, in addition to uncovering certain unsafe 737 aircraft that had previously been "certified" to fly, the FAA may have answered critics who contend that it did not have an accident prevention program.[55] More importantly, however, it signaled a subtle shift from the established regulatory tradition of citing a historically low accident rate as an indication of the safety of the aviation system, to a new pattern of thinking among government regulators about safety. In discussing the issue of aviation safety risk communication, it is clear that *safety is more than the absence of accidents*. What then is "safety?" and how should aviation

---

[*] *See*: British Airways Safety Information System (BASIS). BASIS Seminar, 29th January, 1998, "Developing Flight Data Monitoring Systems." That paper describes the experiences of an airline engineering organization with a changed emphasis from producing data to providing decision support information. *See also:* "Federal Aviation Administration, Office of System Safety - *"Global Analysis Information Network." May 1996*. The GAIN concept paper advanced some ideas for the elements needed to establish an early warning capability for existing and emerging safety concerns.

safety information be communicated with the travelling public? I offer my definition of "safety."

> Safety is the goal of transforming the levels of risk that inheres
> in all human activity.

According to preliminary FAA figures, in 1998, major U.S. airlines flew without the death of a single passenger. For the Air Transport Association's David Fuscus, while acknowledging that constant vigilance was necessary to keep fatalities down, particularly since air travel is up, this, was a "testament to how safe commercial aviation is." For Mary Schiavo, the former DOT Inspector General who has been highly critical of the industry and its regulations, "this should not be seen as a cue for anybody to let up on safety." Flight Safety Foundation president, Stuart Matthews said accidents themselves were rare events and a single year was a poor way to analyze the data.[56] A central thesis of this essay is that safety is more than the absence of accidents and it is a fatal flaw in communicating safety risk to the public to rely solely on sophisticated equations built on historical accident data. Essentially, what's being done is the making of a model about what's likely to happen in the future based on probabilistic studies about what happened in the past. Consider the following: Briefly, on Thursday, May 7, 1998, according to a Press Release (APA 54-98) the FAA ordered inspections and corrective actions that immediately affect an estimated 152 U.S. registered Boeing 737-100 and – 200 models with more than 50,000 hours of flying time. Of the 152 aircraft covered by the order, the following carriers had the largest number of airplanes: United Airlines (44); Southwest Airlines (30); Continental (21); America West Airlines (5); Frontier (5); with the remainder spread among 24 other domestic-registered 737s. Subsequently, on May 10, 1998, in another Press Release (APA 57-98), the FAA extended its initial order for an inspection for both sets of fuel pumps wires in each wing to 737s with between 40,000 and 50,000 flight hours within 14 days. This covered an additional 118 domestic-registered airplanes and 282 worldwide, including U.S. aircraft. There are 1,088 Boeing 737s registered in the United States and 2,716 registered worldwide. Let us examine information downloaded from the *World Wide Web* on the *results* of that historic fleet grounding. According to the information contained in the Press Release, *Status of Boeing 737 Wiring Inspections As of June 16,1998*,[57] the FAA inspections of 500 airplanes revealed that a total of 29 airplanes had less than 50 percent insulation chafing. This condition existed mostly on 27 (unidentified) airplanes with over 50,000 cycles and on 2 airplanes (Continental and United) with 30,000 to 40,000 cycles. There were 9 (unidentified) airplanes (all of which were above 50,000 cycles) found to have exposed wires. The FAA noted that no further information on air carriers was available "at this time." However, "it is not alarmed or unnecessarily surprised at these reports."

A closer look at the publicly available data would suggest that *we might have been lucky.* Had it not been for the proactive grounding, it would only have been a matter of time before the right environmental conditions of temperature and fuel loading would have conspired to produce a massive explosion. While arcing would not occur in wires deeply immersed in fuel, not so in a tank with warm vapors. It was the *forward-looking* identification of this hazardous wiring condition (classic System Safety Engineering) that has averted possible catastrophe. These "certified" airplanes were flying blind since we had no knowledge of these hazards. They were accidents waiting to happen! Subsequent intermittent checks on the FAA's web site (to determine the status of the remaining airplanes inspected) since the July 22nd date when the information was first obtained, revealed that the information had not been updated and made publicly available. Finally, another check on October 1,1998, revealed that that original information (Status Report, June 16, 1998) had been deleted with no reasons offered.

Individuals with a high index of suspicion may choose to question how the FAA went from an initial "not alarmed or unnecessarily surprised at these reports" public statement, to a massive escalation of inspections and ultimately to a planned adoption of new regulations governing fuel system designs and maintenance. Such individuals would naturally suspect that complete information on the inspection results of the historic grounding of May 1998 might have been deliberately withheld from the public to avoid creating alarm or raising questions about the safety of the air transportation system. Conversely, the entire episode could rightly be viewed as a proactive safety posture. Given the nature of our open democracy and the dictates of the *Freedom of Information Act,* it would not take difficult detective work to ascertain the full results of the inspection, including which airlines were affected.

On July 23rd 1998, a new Press Release announced "new measures to reduce potential ignition sources in Boeing 747 center wing tanks."[58] Five months later, in a move deemed an interim action until the determination of the cause of premature wear on some pump shaft bearings, the Agency ordered operators of Boeing 747 aircraft to "immediately change fuel pump procedures to prevent 'dry operation' that could result in ignition of the center fuel or horizontal stabilizer tanks. The Press Release noted that the Airworthiness Directive is "unrelated to the TWA-800 accident, which is still under investigation by the NTSB. The wear conditions were not found on the center wing fuel tank pumps that were recovered from that accident.[59]

The continuing actions by the FAA to ensure fuel tank safety included the ordering of airlines to inspect, within 60 days, fuel boost pump wiring on Boeing 737-100 through-500 series aircraft with 20,000 to 30,000 flight hours. "The inspections are necessary to ensure that the aircraft do not have a problem with chafing and electrical arcing between the fuel boost pump wiring and the surrounding conduit. The directive also requires the addition of a layer of Teflon sleeving to protect the fuel pump wires."[60] The FAA estimates that 215 U.S.-registered 737s, with 20,000 to 30,000 total flight hours, would be affected. The aircraft are operated by most major U.S. airlines. The Press Release pointed out that the Boeing 737 models—the 737-100-600, and –800 do not have electrical wires running through conduits within fuel tanks. Finally, on October 1,1998, Secretary of Transportation Rodney E. Slater and FAA Administrator Jane F. Garvey announced a multi-year effort—which includes both short-and long-term initiatives—to address the safety and reliability of systems on commercial aircraft. The Press Release went on to state that regulations will be proposed "to require certain aircraft manufacturers to demonstrate that fuel system designs remain safe and prevent possible ignition sources in the fuel tank."[61]

The above material was cited to support the case for *improved risk communication* and a major thesis of this book that *safety is more than the absence of accidents.* The previously cited text on risk communication, *Mad Cows,* provided detailed analyses of risk communication practice and malpractice and provided a set of ten lessons for risk management communicators. To single out a few: *Lesson 2: Regulators are responsible for effective risk communication.* Agencies of governments having regulatory authority over a broad range of health and environmental risks have or are capable of acquiring through legislation, the legal authority to manage risks. This means that they can devise and enforce rules that can modify behavior across the entire range of risk situations including in the workplace and on the roads and highways to name a few. "In general, the objectives of these agencies are risk control and reduction, not risk elimination, because as a general rule risk—a certain probability of harm—is an inescapable part of activity in the environment."[62] *Lesson 3: Industry is responsible for effective risk communication.* Industry is in an unusual position with respect to the hazards its products and processes entail. In the above cited aviation case material, the statement that the FAA it is not alarmed or unnecessarily surprised at these reports" appeared to be a violation of *Lesson 8: banish "no risk" messages.* The authors, Powell and Leiss, noted that although citizens and environmentalists are often taken to task by government and industry for advocating "zero risk" scenarios, pronouncements of the "there is no risk" variety are a favorite of government officials. "A zero risk policy is the functional equivalent of exorcism."[63]

Safety Data vs. Information

For advocates of greater available information to consumers of air transportation services, one of the dominant issues is public knowledge of airline safety. How do air travel consumers *know* that airline X, Y, Z is safe? And what are the available data and information resources that can assist consumers with decision making on airline choice? Currently, a plethora of publicly available aviation safety risk data exists on the Internet provided by multiple sources including the FAA, the NTSB and the Bureau of Transportation Statistics.

Based on the safety statistics, airline accidents are extremely rare events. The airspace system *daily* handles more than 174,000 takeoffs and landings at airports across the nation, and *routinely* carries approximately 1.7 million passengers *safely* to their destinations. In fact, according to MIT's Dr. Barnett, under 1987-96 data, a passenger who took a First World domestic jet flight every day would on average go for 21,000 years before succumbing to a fatal crash. The system is so safe that "discriminating among airlines to improve survival odds is a fruitless pursuit. ...there is no way to exploit statistical differences in risk to improve our safety on a given trip."[64] Barnett further notes that the crash of Swissair Flight 111 in September 1998 off the coast of Halifax was especially jarring because the airline is "the embodiment of aviation safety...while any of us could die in an air crash, almost none of us will. There is no point in trying to beat the odds."[65] Yet, with every occurrence of one of those rare events, the media interviews the usual "safety experts," airline safety advocacy individuals and groups who predictably voice concerns over airline safety and call for more stringent regulatory oversight along with the making of safety information available to aviation consumers for intelligent decision making on airline choice. In fact, in 1996, the Wyden/Ford initiative urged the FAA to take action to educate the public better about the safety of the aviation system and make important information about aviation safety more easily available to consumers. The FAA responded positively and, among other items, initiated a quarterly compilation of all FAA enforcement actions against regulated aviation entities that involve safety and security issues.

The issue of decision making on airline choice is fraught with peril. For consumer advocates, the decision is often naïvely stated in terms reminiscent of a consumer's decision to purchase a particular automobile over another. From my perspective, while there may be no way for the aviation consumer to statistically exploit differences in risk to improve one's perceived safety, there remains the fundamental issue of the inherent right of the consumer* to information about the imposition and/or mitigation of safety risks by profit-making firms. For example, basic information such as the existence or lack of a company's Safety Policy would include the title and rank of an airline's safety officer and key elements of the safety program, the nature and scope of the "Safety Organization" and participation in safety training programs. For example, information about management policy among airlines that requires mechanics to check out their tools (a process implemented by Boeing in 1998) at the start of each shift would serve to underscore an aspect of the company's safety culture. Naturally, such a practice is no guarantee against the occurrence of an accident. However, effective risk communication of a company's safety culture would partially fulfill what Baruch Fischhoff of the

---

* In early 1999, the issue of an Aviation Consumer Bill of Rights received congressional attention. Hearings were triggered by an incident—a snowstorm that left hundreds of Northwest Airlines passengers stranded for up to eight hours aboard aircraft in Detroit. In June 1999, the airlines and congress jointly announced a "Customers First" program in lieu of the proposed legislation, to address frequent customer complaints, such as delays/cancellations and nightmarish encounters with airline bureaucracies. Months later, Paul Hudson of the Aviation Consumer Action Project commented that "something's not right when the chief sponsors abandon a reform bill without consulting any consumer groups. ... This wasn't just a sweetheart deal; it was a giveaway."

Department of Engineering and Public Policy, Carnegie Mellon University, called the "social contract between those who create risks and those who bear them."[66]

The FAA, NTSB and BTS have made volumes of statistical data including the entire Federal Aviation Regulations (FARs) available on the Internet. However, there is no declarative policy statement as to the purpose of providing such data. Even assuming one had the time, the biggest problem for the consumer is to be able to intelligently sort through the deluge of documents, regulations, maps and raw data that has been indiscriminately dumped on the Internet. The fact that the data is not packaged and formatted in a meaningful and readily understood information format for the benefit of consumers might have the unintended effect of undermining aviation safety risk communication efforts. For e.g., various newspapers provide a so-called "safety ranking" of airlines based solely on the number of accidents from the NTSB database. Besides the inherently flawed logic of a consumer relying solely on the number of accidents by airline X, Y, or Z, as an indicator of airline safety, this incomplete information may unnecessarily alarm the traveling public and inflict economic harm on the airlines. Not surprisingly, lawsuits have ensued.

Clearly the issue of airline safety is well beyond the usual citations of a "very low accident rate" as proof positive of a safe system. It is also beyond "Black Box" technology consensus building from the viewpoint of understanding the fundamental philosophical underpinnings of the relationship between the regulated and the regulators; and the relationship between the degree of government oversight and independence. While the communication of aviation safety risk information ought to be a cooperative endeavor among aircraft manufacturers, airline management, regulatory agencies, industry associations, pilots and interest groups, such cooperation should also include the sharing of strategic safety planning initiatives among the regulated and the regulators. It is not enough that aviation industry executives be invited to voice their views on the FAA's Strategic Plan while a reciprocal arrangement does not exist for information on industry strategic plans that impact safety. By its nature, aviation safety is a cooperative effort that transcends the notion of relying solely on regulatory policing and enforcement. Realistically, how can the regulators be expected to develop sensible strategic plans without specific clues as to what the regulated are planning, particularly in terms of addressing projected traffic increases and the concomitant risk mitigation strategies?

Clearly, this complex subject can be made more understandable to the public through industry-government cooperation and the development of a framework for aviation safety risk communication to include the following:

- The lack of immunity from regulatory prosecution for those who voluntarily provide safety operational data creates heightened mistrust and remains a major roadblock to the sharing of safety related information.
- The lack of a declarative policy statement on the provision of aviation safety consumer information has created a risk communication vacuum that exacerbates safety risk controversies with every occurrence of a rare accident.
- Risk communication is the process of exchanges about the steps taken on the identification, characterization and mitigation of risks to reduce public anxiety. Effective risk communication of a company's safety culture would partially fulfill the "social contract between those who create risks and those who bear them."
- The provision of "accident data" on the Internet in a manner not easily understood by consumers may invite the potential for its misuse and the unintended effects of undermining aviation safety risk communication efforts.
- Safety is more than the absence of accidents. It is a fatal flaw to attempt to communicate future risk likelihood based *solely* on studies of past experience.
- Safety is the goal of transforming the levels of risk that inheres in all human activity.

To further underscore the idea that *safety is more than the absence of accidents,* let us briefly consider the following regulatory actions. On July 14,1998, the FAA announced a $5 million civil penalty with America West Airlines of Phoenix, Arizona to settle alleged violations of aircraft maintenance and operations regulations. The airline which had a 1997 reported revenue of $1.9 billion and a profit of $75 million, had flown more than 41,000 hours over two years in violation of Federal Aviation Regulations and ignored at least some of the regulations. The FAA cited 17 Airbus A320s, which flew without visual structural inspections which the manufacturer, Airbus Industrie, had recommended in 1994; eight Boeing 737s, which were missing cargo netting. Such netting is designed to prevent cargo from shifting in flight; and one Boeing 757, which was missing an elevator actuator. The elevator, attached to a plane's horizontal stabilizer, controls the jetliner's ascent and descent movements. America West pointed out that although one of the actuators for the tail-flap mechanism was not working, the backup was. For Mary Schiavo, the "airline has a good safety record. It ranks second best, behind Southwest Airlines, for its rate of accidents and incidents. However, redundancy is the key to airline safety, and without a backup, the aircraft was at risk. Since the tail allows a plane to climb, descend and maintain flight. The fine is evidence of a tougher FAA."[67]

The significance of the historical May and July 1998 regulatory actions cited above lies in the fact that they represented a strategic departure from post accident investigation to accident *prevention and avoidance.* As a result of those sweeping regulatory actions, some airlines with the "best safety records" over a decade were found to be at risk. Clearly, the reliance on past accident data, while a necessary condition, is not sufficient for determining and communicating "safety." The lack of systematic knowledge about hazards and their risk mitigation strategy lends credence to the belief that it is entirely possible to not have an accident in decades yet be operating an unsafe system. We can only say definitively that a complex system such as the National Airspace System is safe once we have identified the hazards and adopted a risk mitigation strategy. Until this is done, unknown, unidentified hazards could someday turn into a major catastrophe. Therefore, it is these *forward-looking* regulatory actions, more than any other that enables us to answer the question: How do we know that a system is safe? Or how can a system be made safer? A 1999 Rand Corporation report, *Safety In The Skies,* recommended that the NTSB should take a proactive stance and report safety problems and trends before they lead to serious accidents.

Design-Induced Errors

U.S. and European basic fail-safe design concept is that: "no single failure or probable combination of failures during any one flight shall jeopardize the continued safe flight and landing of the airplane." This and other safety considerations such as crashworthiness requirements for the entire airplane, aircraft performance requirements after loss of engine power etc have been incorporated into the *Federal Aviation Regulations (FARs).* For an understanding of the early underlying safety assessment concepts of aircraft systems in the U.S. and European regulations for transport aircraft, the book, *Systematic Safety* by E. Lloyd and W. Tye is required reading. The foreword of the book notes that the underlying philosophy or basic approach to achieving safety objectives are only rarely issued by the U.K. Civil Aviation Authority (CAA) and do not fall easily into any of the established publications.[68]   In the 1998-reprinted version, the authors cautioned that developments in technology have caused some areas of the text to be out-dated.   Nonetheless, the book provides the basis for understanding the aircraft safety requirement that an inverse relationship should exist between the probability of an occurrence and the degree of hazard.   Many of these requirements are the result of knowledge gained from accident investigations.   "Back in the 1970s and 1980s, the fatal accident rate was quoted at 0.8 x 10-6 per hour.   Also, the fatal accident rate for the 'best' aircraft types was in the order of 10 times better than the 'worst' types.   However, taking the total figure and allocating shares to the different causes of accidents, a large portion is attributed to pilot error, some to the weather and about 10 percent are attributed to aircraft systems failure.   Hence, the portion attributed to all aircraft systems is in the order of 1 x 10-7 per hour (or less).   If we assume as many as 100 individual, independent potential catastrophic systems events are uncovered for a typical modern aircraft then each event must be designed to the less probable than 10-9 per hour.   Looked at in another way, consider a fleet of 100 aircraft of a type, each one flying for 3000 hours per year.   Annually, that's 300,000 hours.   If the aircraft type is in service for 30 years, then 9 million hours are accumulated.   Thus, a single catastrophic systems event should virtually never happen if the safety target for aircraft systems is set at 10-9 per hour."[69]

In the 1990s, the aviation industry and government cooperatively developed the voluminous *Aerospace Recommended Practices (ARPs) 4754 and 4761* as a set of "best practices" for the development of new systems and an improved aerospace safety process. The ARPs introduced the concept of preliminary safety assessment (PSSA) for systems on civil aircraft.   In January 1999, at an FAA sponsored seminar on Safety Risk Assessments, several members of the aerospace industry stated that the FAA should focus on those standards and guidance material that encompasses a total "systems" approach to include human factors.   It was pointed out that unlike *Military Standard 882c*, System Safety Program Requirements, ARP 4761, Guidelines & Methods for Conducting The

Safety Assessment Process on Civil Airborne Systems and Equipment, does not cover the full realm of the system.[70] At the University of York in the United Kingdom, researchers with experience in working with aircraft manufacturers in Europe and system suppliers in Europe and the U.S. have illustrated some "key difficulties" with the ARPs including: the tendency to mistake the role of the PSSA as proof positive that "the proposed design *is* safe, not that it *can be* safe *if* the components are implemented appropriately." The PSSAs have a crucial role in contributing to the design, hence, there is the need for more closely integrated working of the design and safety teams. The Guidelines do not address the issue of human factors, which is "arguably the most significant factor in aviation safety."[71]

As mentioned in the introduction, the thesis of Steven M. Casey's book, *Set Phasers On Stun And Other True Tales of Design, Technology, and Human Error,* is that many "human error" accidents should be more aptly named design-induced errors. In modern times, the most startling evidence of a designed-induced error in aviation was the Airbus A320. The following paragraphs have been summarized from Dr. Casey's book.[72] That case study demonstrates how far we are from adequately controlling hazards and managing risks. At the outset, the lack of any publicly available material from the aircraft manufacturer's perspective is a communication failure. *Mention should also be made of the fact that all airline operators routinely report service problems to their manufacturers and there is no publicly available evidence to suggest the safety of one aircraft type over another.*

Essentially, the Airbus is a *fly by wire* aircraft. Such a system uses shielded electrical lines, instead of metal cables or vulnerable hydraulic tubes, from the cockpit controls to the control mechanisms of the wings and engines. The electrical cables from the instruments feed into a bank of five computers. The computers analyze the inputs received from the pilot, and compare them to various flight parameters such as speed and altitude for "flight protection." *After* this analysis, it sends the signals to the wings, engines, or other major components for execution. The idea of a computerized "flight protection" was deemed as a *revolutionary safety feature* for commercial aircraft. The pilot would not be permitted to make any unintentional errors. The computers, programmed with the laws of flight and the infinite characteristics of the aircraft, were capable of directly intervening to avoid an unintentional error that for e.g. would induced a stall by going too slowly or pitching the aircraft up too high. Also, the pilot would not be permitted to perform any actions that exceeded the forces the airframe was designed to withstand, the computers would intervene and adjust the pilot's command and kept the aircraft within the limits of structural safety.   According to Dr. Casey, within Airbus, the flight protection system was seen as analogous to a guardrail on a mountain road. If the driver lost control of the vehicle, the guardrail prevented the driver from going over the edge.   The computers on the A320 kept the plane both within the flight envelope and its structural limits.   For Dr. Casey, the design was a "textbook case" in the use of a computer to monitor and keep the human operator out of trouble.

During the low-level flyover demonstration at the air show at Mulhouse Habsheim in the Alsace region of France, on June 26,1988, the captain disengaged the *autothrottle* and the *alpha floor* function.   The autothrottle is similar to a cruise control on an automobile.   On the airplane, it automatically maintained a specified speed regardless of the plane's position.   The alpha floor flight protection kept the plane from stalling during approaches and takeoffs due to any combination of low speed, angle of attack, or loss of lift.   The flight protection system activated automatically when the plane pitched upward more than 15 degrees and its altitude was 100 feet or more above the ground.   The flight plan called for a steep climb up that would exceed 15 degrees.   Therefore, to prevent the computer from prohibiting the steep upward climb, the pilot deactivated the automatic protection system.   The alpha floor stall protection did not automatically activate at an altitude below 100 feet.   (Pilots land planes by reducing speed on the final approach and then flare the wings just as they are over the runway, inducing a controlled stall.   With the loss of lift, the plane settles gently on the runway).   The alpha floor protection system safeguarded the plane from stalling.   Therefore unless it was automatically turned off at a very low altitude the plane never would be "allowed" to land, resulting in the crash at the treetops.   It was determined that the very system designed to increase safety of the aircraft, the flight protection system, inspired overconfidence in the captain and was a probable cause of the accident.   The pilot tried to perform a maneuver that was well beyond the limits of safety.   Instead of treating the flight protection system as the safety guardrail on the mountainous road, the pilot used it to define the upper limits of aircraft flight.   It was as if the pilot "used the guardrail to negotiate the curve, not ...as a protective barrier placed there in the event (of) lost control."   The pilot had turned the flight protection system off for the maneuver and had to fly by the standard rules of flight and may not have been prepared for this situation.

On February 14, 1990, 19 months after the Mulhouse airport accident, an Indian Airlines Airbus A320 crashed on approach at Bangalore Airport, India. The accident investigators determined that a disconnected autothrottle and selected *idle open descent mode* during the final phase of a descent to the airport. "The system designed to be used only at high altitude during descent with reduction in speed, was activated during approach. Unknown by the crew, the engines slowed to idle speed and the plane, flying at a minimum speed and maximum angle of attack, began to stall at an altitude of 140 feet. The application of full engine throttle to overcome the inertia of the idle engines was not timely and the aircraft crashed. The accident investigators concluded that crew overconfidence played a major role in the accident. In March 1990, Airbus Industrie conducted a special A320 operators meeting on the topic of "overconfidence syndrome." Dr. Casey noted that "In addition to changing the training strategy, plans to upgrade the alpha floor protection mode into an approach protection mode to prevent descents without sufficient thrust" were discussed.

At this point, more than a passing reference should be made to the FAA fail-safe design concept guiding the Part 25 airworthiness standards and fully defined in Advisory Circular 25.1309-1A. The fail safe design concept considers the effects of failures and combinations of failures in defining a safe design. Essentially, the design concept uses a combination of two or more design principles (e.g. designed integrity and quality) or techniques in order to ensure a safe design:

- In any system or subsystem, the failure of any single element, component, or connection during any one flight should be assumed, regardless of probability. Such single failures should not prevent continued safe flight and landing, or significantly reduce the capability of the airplane or the ability of the crew to cope with the resulting failure conditions.
- Subsequent failures during the same flight, whether detected or latent, and combinations thereof, should also be assumed, unless their joint probability with the first failure is shown to be extremely improbable.

Although the extensive operating experience gained in several decades of flying has made the aircraft system very safe, as can be seen from *Table 3:2*, the world-wide commercial jet fleet has had accidents in which design related issues were the primary causal factors.

Table 3:2 Design-Related Accidents: 1959 - 1996

System	Worldwide		U.S. Operators	
	Total	Percentage	Total	Percentage
Power plant or thrust reversers	16	2.7	5	2.4
Landing gear, brakes and tires	13	2.2	3	1.4
Flight controls	9	1.5	2	.9
Electrical systems and instruments	6	1.0	3	1.4
Hydraulics	2	.34	1	.48
Passenger accommodations	2	.34	0	0.0
Auxiliary power	1	.17	1	.48
Fuel systems	1	.17	1	.48
Total Airplane Systems	50	8.67	16	7.8
Total All Causes	577		205	

Source: Air Transport Association

The purpose of including this data is to provide a context for thinking about the thesis of this essay that safety is more than the absence of accidents. Dr. Casey has succinctly summed up the dilemma with the statement that: Traditional engineering knowledge—of electronics, chemistry, physics, structures, and materials—is insufficient in and of itself for the design of technologies which play such a profound role in our lives. The future will require more of designers and their designs as systems become more complex, more intertwined, and even more, not less, dependent on human capabilities and limitations. The data also provides a context for thinking about the historic *accident prevention* regulatory actions of 1998 involving fuel systems. For certain, if a regulatory philosophy had been based on for example a Pareto analysis of past accidents, efforts would certainly not have been directed to the relatively very few accidents in the fuel systems area.

Concluding Remarks

This Chapter has attempted to guide readers through selected areas of the vast literature of reliability engineering and risk management that have narrowed the gap between the fairly esoteric theories of reliability mathematics, risk management analytical techniques, communications and understanding the role of the human in a system:

- Dr. Richard B. Jones, author of *Risk-Based Management, A Reliability-Centered Approach,* has observed that most maintenance and operations personnel view the risk models that have been developed around rigorous mathematics, probability, and statistics as not being useful to meeting the challenge of guaranteed reliability at an affordable price.   The Jones text examined the ideas and methods that have been used to improve reliability measurement, and reduce risk. The text by Bertram Amstadter, *Reliability Mathematics,* served the needs of the undergraduate or practitioner who has had only an introductory course in statistics/statistics quality control.   In a single ready reference, this applied rather than theoretical text provides useful tools for addressing reliability statistics problems.

- There is ample empirical evidence that demonstrates that certain operations with impeccably reliable and accident-free records have in fact been unsafe. And it would only be a matter of time before "the dynamics of accident causation" would set.   The text, *Human Error,* by Dr. James Reason, contains a wide discussion of the trajectory of accident opportunity penetrating several defensive systems. This approach would caution safety practitioners to guard against the propensity to believe that a reliable system is a safe system.   The communication of aviation safety risk information ought to be a cooperative endeavor among airline management, regulatory agencies, industry associations, pilots and interest groups. The book, *Mad Cows And Mother's Milk,* contains instructive lessons in communicating risk to the public. It has provided seven case studies on risk communication failures between scientific experts and the public.

- The National Research Council's work of the Panel on Human Factors in Air Traffic Control Automation, *Flight To the Future,* emphasized the point that designers strive to meet specific FAA procurement requirements (for e.g., build to 3 nines). Whereas this appears to be an admirable goal, absolute reliance on specified reliability should be treated with some caution for three reasons: (1) Like any estimate, a reliability number has both an expected value (mean) and an estimated variance. The variance is often ill defined and hard to estimate. When it is left unstated, it is tempting to read the offered reliability figure (e.g., r = .999) as a firm promise rather than the midpoint of a range. (2) Objective data of past system performance reveal ample evidence of systems whose promised level of reliability greatly overestimated the actual reliabilities. (3) Experience also reveals that it is next to impossible to forecast all inevitable circumstances that may lead a well-designed system to "fail," even given the near boundless creativity of the system engineer.

- Nicholas J. Bahr's text, *System Safety Engineering and Risk Assessment: A Practical Approach,* and the Stephenson text, *System Safety 2000,* contains valuable tutorials on fault tree analysis. Bahr cites a few mistakes that are to be avoided in constructing, quantifying, and evaluating fault trees. What is important to remember is that fault tree analysis is not a model of all possible system failures or all possible causes but rather, a model of particular system failure modes and their constituent faults that lead to the accident. The award winning International System Safety Society conference paper: "Beauty And The Beast – Use And Abuse Of The Fault Tree As A Tool" should be required reading for all safety analysts. The paper, presented by the System Safety Engineer, R. Allen Long, described proper application and misapplications of the fault tree as a tool when evaluating complex systems.

Notes

[1] Martin L. Shooman, *Probabilistic Reliability. An Engineering Approach* (New York: McGraw-Hill, Inc., 1968.

[2] Henry Petroski, *To Engineer Is Human,* (New York: First Vintage Books Edition, 1992), pp. vii-viii.

[3] IBID., pp. 98 – 100 *passim*

[4] IBID., pp. 163.

[5] Scott D. Sagan, *The Limits of Safety: Organizations, Accidents and Nuclear Weapons,* (Princeton, NJ: Princeton University Press, 1993), pp.13 –17 *passim.*

[6] M. .Modarres, *Reliability And Risk Analysis,* (New York: Marcel Dekker, Inc., 1993), p. 5.

[7] Nancy Leveson, Op. Cit., p.171.

[8] The following paragraphs are paraphrased from Martin L. Shooman, pp. X, 1.

[9] Paraphrased from Shooman, Op.Cit., p.119.

[10] *The Washington Post,* July 29, 1998, "Lois Gibbs's Grass-Roots Garden," p. D.1.

[11] John D. Graham, Laura C. Green, and Marc J. Roberts, *In Search of Safety,* (Cambridge,MA: Harvard University Press, 1988).

[12] National Research Council, *Science And Judgment In Risk Assessment,* (Washington, DC: National Academy Press, 1994), Prepublication Copy, pp. E4 - E5.

[13] Perrow, Op.Cit., pp. 304-310 *passim.*

[14] Vernon L. Grose, *Managing Risk,* (Arlington, VA: Omega Systems Group, 1987), pp. v-vi.

[15] IBID., p. 89.

[16] Hammer, *Handbook of System and Product Safety,* p. 31.

[17] David Gero, *Aviation Disasters,* (London: Butler & Tanner Ltd., 1996), 2nd Ed.

[18] M. Modarres, *Reliability And Risk Analysis,* (New York: Marcel Dekker, Inc., 1993), p. 7.

[19] Peter L. Bernstein, *Against The Gods,* (New York: John Wiley & Sons, Inc. 1996).

[20] National Research Council, *Flight To The Future,* (Washington, DC: The National Academy Press, 1997), pp. 18-19.

[21] N.H. Roberts, W.E. Vesely, D.F. Haasl, and F.F. Goldberg, *Fault Tree Handbook,* NUREG-0492, (U.S.Nuclear Regulatory Commission, Washington, DC, 1981).

[22] Paraphrased from Bahr, Op. Cit., pp. 127-128.

[23] R. Allen Long, Senior System Safety Engineer, Hernandez Engineering, Inc., "Beauty And The Beast – Use And Abuse Of The Fault Tree As A Tool 17th International System Safety Conference *Proceedings,* pp. 117- 127.

[24] *Systematic Safety,* (London, U.K. 1982: Civil Aviation Authority).p. 3.

[25] Reason, Op. Cit., pp. 218-233 *passim.*

[26] Probabilistic Safety Assessment And Management (PSAM 4) *Conference Proceedings,* "HuRa! – A Prototype Expert System for Quantifying Human Errors," Vol. 1, p. 536.

[27] PSAM 4 *Conference Proceedings,* Elisabeth Pate-Cornell, "Priorities in Risk Management: Human and Organizational Factors as External Events and a Maritime Illustration," Vol. 4, p. 2675.

[28] IBID., p. 21.

[29] Leveson, Op. Cit., p. 8.

[30] Kishor S. Trivedi, *Probability and Statistics with Reliability, Queuing, and Computer Science Applications,* (New Delhi: Prentice-Hall of India, Private Limited, 1997), pp. 30-32.

[31] Sagan, Op. Cit., p. 28.

[32] Perrow, Op.Cit.p.78.

[33] Sagan, Op. Cit., pp. 28-44 *passim.*

[34] Sagan, Op. Cit., p.46.

[35] "Spilled Coffee Led To Near Miss," *The New York Daily News,* 6/28/98; "Blame for Jets' Near-Collision," *Newsday,* 6/29/98; "Coffee Spill Caused Close Call at LaGuardia," *The Washington Post,* 6/28/98, P.A7; "Cause Suggested for Near Miss," *The New York Times,* 6/29/98, p. A17.

[36] CNN-FAA Investigating Close Call with 767 and Concorde at JFK-May 25,1999.

[37] *ABC NEWS,* "Construction Worker Zaps Miami Air Control Center." http://abcnews.go.com/wire/world/reuters19990709.

[38] *ABCNEWS.com*: Concordes Pass too Close over JFK-Aug 27,1999.

[39] *The Times,* August 28th 1999, p. 8.
[40] Adapted from, Grover Starling, *The Politics and Economics of Public Policy,* (Homewood, IL: The Dorsey Press, 1979), pp. 60-61.
[41] *The New York Times,* Editorial, "What's Wrong With Air Travel," June 19,1989.
[42] *The Wall Street Journal,* "Overbooked: A Crush of Air Traffic, Control-System Quirks Jam the Flight Lanes." September 1, 1999, p. 1.
[43] *Aviation Daily,* "Team Effort Needed To Resolve Aviation's Problems," September 22, 1999.
[44] "Air Travel in Crisis," *Aviation Week & Space Technology,* October 25, 1999.
[45] "Canada's Private Control Towers." *The New York Times,* October 23, 1999.
[46] "Pilots: Runway Crossings A Safety Hazard," *USA Today,* November 13, 1998, p.1.
[47] The Royal Society, *Risk. Analysis, Perception and Management.* (London: The Royal Society, 1992), p. 118.
[48] IBID., p. 89.
[49] Douglas Powell and William Leiss, *Mad Cows And Mother's Milk,* (Montreal & Kingston, Canada: McGill-Queen's University Press, 1997).
[50] IBID., p. 33.
[51] 1998 Transportation Hearings before a subcommittee of the Committee on Appropriations, House of Representatives. One Hundred Fifth Congress, 2nd session, Part 6, p.231.
[52] Gero, Op. Cit. *Aviation Disasters.* pp. 206-208.
[53] "Boeing Delayed Fuel Tank Report," *The Washington Post,* October 30, 1999, p. 1.
[54] "FAA Orders Immediate Inspection for High-Time Boeing 737's, Extends Inspection Order," FAA Office of Public Affairs, APA 57-98, May 10, 1998.
[55] Letter from C.O. Miller to FAA Administrator, Jane Garvey, October 8th, 1997.
[56] *CNN,* "No U.S. Airline Deaths in '98," January 6,1999.
[57] "Status of Boeing 737 Wiring Inspections As of June 16,1998," downloaded from the Web on 7/22/98: www.faa.gov/apa/737iu.htm.
[58] *FAA News,* "FAA Acts To Increase Center Fuel Tank Safety," APA 92-98, July 23,1998.
[59] *FAA News,* "FAA Issues Emergency Order on Boeing 747 Fuel Pumps," APA 143-98, December 3,1998.
[60] *FAA News,* "FAA Continues Boeing 737 Wiring Inspections," APA 117-98, Sept. 28,1998.
[61] *FAA News,* "FAA Unveils Plan To Enhance safety of Aging Aircraft Systems," DOT 180-98, October 1, 1998.
[62] Douglas Powell and William Leiss, *Mad Cows And Mother's Milk,* Op.Cit., pp. 215-216.
[63] IBID., PP.223.
[64] "Flying? No Point in Trying to Beat the Odds," *The Wall Street Journal,* Sept. 9,1998, p. A22.
[65] IBID.
[66] Baruch Fischhoff,"Risk Perception and Communication Unplugged: Twenty Years of Process." *Risk Analysis,* Vol.15. No.2, 1995, pp. 137-143.
[67] "Record Safety Fine For Am West," *The Arizona Republic,* July 15,1998; also"FAA Announces Civil Penalty Settlement With America West," *FAA News,* APA 90-98, July 14,1998.

[68] *Systematic Safety,* (London, U.K. 1982: Civil Aviation Authority).

[69] "Where did the ten to the minus nine come from?" Notes by John Vincent, U.K. Civil Aviation Authority. Meeting of RTCA SC-189/EUROCAE-WG53, Toulouse, FR. February 8th, 1999.

[70] "Industry to FAA: Take A Systems Approach To Safety Assessments." *FAA Safety Risk Assessment News,* Report No: 99-1, January/February '99.

[71] S.K. Dawkins, T.P. Kelly, J.A. McDermid, J. Murdoch, D.J. Pumfrey, "Issues In The Conduct of PSSA." 17th International System Safety Conference *Proceedings,* pp. 77- 86.

[72] *Set Phasers On Stun, op cit.,* 92-106 *passim.*

4 System Safety Engineering School

"A popular misconception is that by eliminating failure a product will be safe. A product may be made *safer* by eliminating or minimizing failures, but there are other causes of accidents…mishaps often occur where there is no failure."

Willie Hammer, PE

Reliability versus Safety

To continue with the classification framework, whereas in the disciplines we separated things, in *synthesis,* we put things together. Readers are advised that there are some unavoidable overlaps with other chapters that may conflict with the order of theoretical presentation.

No discussion of the System Safety discipline can be conducted without reference to Willie Hammer's authoritative text, *Handbook of System and Product Safety.*[1] Many authors have been inspired by his work. Hammer's prescient observation that it is a misconception to assume that by eliminating failures, a product will be safe, set the stage for the debate of truly seismic proportions of reliability versus safety. In the preceding chapter we have examined the fundamental tenets of the "Reliability Engineering School" and its equation of "reliability" with "safety." The importance of Hammer's works lies in the fact that he recognized that there are other causes of accidents: dangerous characteristics of the product, human action, extraordinary environmental factors, or combinations of these. The 1970 *Final Report of the National Commission on Product Safety* discussed numerous products that have been injurious because of such deficiencies. The majority of the injuries stemmed from the results of hazardous characteristics rather than failures. He also noted that in the liability lawsuits attributed to the hazardous performance of the Corvair, the claimants cited dangerous characteristics due to negligent design, not mechanical failure.[2]

99

Hammer's text has documented the enormous losses of military aircraft and pilots and the perspective of designers in developing emergency procedures and equipment to be used when failures occurred. This overdependence on the pilot's ability precluded earlier application of accident prevention principles to the elimination of engineering deficiencies. With the development of the ballistic missiles, it became obvious that the problem of accidents lay in the design and production of the missile. It now became apparent that many safety problems could be solved only by good design. For Hammer, the system safety (and product safety) concept is predicated on this principle:

> The most effective means to avoid accidents during system operation is by eliminating or reducing hazards and dangers during design and development.

What Is System Safety

The modern discipline of System Safety evolved in 1962, with the dawn of the space transportation era. It was not until the early 1960s that the system safety concept as formally applied by contractual direction, with the April 1962 U.S. Air Force system safety requirements on the Minuteman missile program. With minor revisions, the document, "System Safety Engineering for the Development of the Air Force Ballistic Missiles" evolved into a Defense Department mandatory requirement (MIL-STD-882) on all procured products and systems. This formal delegation of safety responsibility by contractual requirement replaced the familiar practice in which each designer, manager, and engineer presumably assumed their share of the responsibility for safety. The growth and development of the System Safety approach was strengthened by the publication of safety standards, specifications, and requirements, as well as operating instructions. The contractor is required to establish and maintain a system safety program to support efficient and effective achievement of overall system safety objectives. A system safety program consists of the following: (1) System safety program plan; (2) Identification of hazards; (3) Assess mishap risk; (4) Identification of risk mitigation measures; (5) Reduce risk to an acceptable level; (6) Review and acceptance of residual risk by approval authority; (7) Communicate to the approval authority a list of identified hazards, corrective actions incorporated in the system, and known residual risk in the system. By the 1990s, the Department of

Defense acquisition reform process has focussed on the elimination of detailed specs and standards; however, both the Defense Department and industry have realized the unique role that MIL-STD-882 has played in the design of safe products. While it could not be eliminated, to facilitate the system safety process, it would have to directly address performance and objectives as opposed to merely specifying tasks. A new performance-oriented approach is addressed in Revision D of MIL-STD-882.

According to Roland and Moriarty, authors of, *System Safety Engineering And Management,* this 1990 edition of their earlier 1983 textbook is intended to serve as a reference manual for the practicing System Safety professional. The objectives of the discipline of System Safety are to impose a state of safety on systems under design. This includes evaluating the system, determining if that state of safety is acceptable, evaluating safety countermeasures, and making decisions concerning the employment of a set of countermeasures and the use of the system. System Safety is defined as the application of special technical and managerial skills to the systematic, *forward-looking* identification and control of hazards throughout the life cycle of a project, program, or activity. The concept calls for safety analyses and hazard control actions, beginning with the conceptual phase of a system and continuing through the design, production, testing and disposal phases, until the activity is retired.[3]

A system has been defined as a composite of people, procedures, materials, tools, equipment, facilities, and software operating in a specific environment to perform a specific task or achieve a specific purpose, support, or mission requirement. System Safety is the assurance that the system in question is safe. Hazard analysis is the heart of the system safety approach. Anticipating and controlling hazards at the design stages is the key to a System Safety approach. For them, "System Safety is *not* failure analysis. Hazard is a much broader term than failure." The revised 882D defined a hazard as *"any real or potential condition that can cause injury, illness, or death to personnel; damage to or loss of a system, equipment or property; or damage to the environment."* Roland and Moriarty reinforced Hammer's earlier view on failure by pointing out that a *failure* is when something functions in a manner in which it was not intended. A failure can occur without loss. Severe accidents have happened while something was operating exactly as intended—which is without failure. Professor Leveson's *Safeware: System Safety and Computers,* has devoted extensive sections to the System Safety Engineering discipline and incorporates concepts from the earlier works of many System Safety theoreticians and practitioners. The title of the book expresses the "impossibility of separating the various aspects of the system in dealing with safety issues. The approach requires a change in attitude and changes in design and development processes."[4]

The 1999 text by Felix Redmill, Morris Chudleigh and James Catmur, *System Safety: HAZOP and Software HAZOP,* describes the use of the hazard and operability (HAZOP) technique to the identification and analysis of hazards in software-based systems. For many years, HAZOP has been the foremost hazard identification technique in use by the process (chemical, pharmaceutical, oil and gas) industries. The text also includes a chapter (HAZOP of Human-centered Systems) that describes an approach to the identification of hazards in human-centered systems, in which the human component is central or critical to system functionality or safety. According to the authors, the approach—Safety analysis of User System Interaction (SUSI) was developed by Cambridge Consultants Limited and Arthur D. Little.[5]

System Safety Objectives

The foremost general objective of System Safety is to prevent accidents. Generally, this is achieved by defining safety requirements for the Managing Authority to provide control and technical surveillance. Included, but not limited to, are the following general activities:

- Identify, eliminate, or control hazards.
- Provide maximum designed-in safety consistent with requirements.
- Minimize personal injury, material failures, and monetary losses.
- Minimize the need for safety driven system modifications.
- Verify that normal operation of a system cannot degrade the safe operation of another.
- Protect the public.
- Document actions to reduce risk.
- Document safety data for "lessons learned" exercises.

System Safety Program

A System Safety Program consists of the following:

- System Safety Program Plan.
- Identification of hazards.
- Assess mishap risk.
- Identification of risk mitigation measures.
- Reduce risk to an acceptable level.
- Review and acceptance of residual risk by approval authority.
- Communicate to the approval authority a list of identified hazards, corrective action (s) incorporated in the system, and known residual risk in the system.

Human, Organizational and Environmental Factors

Human engineering has been defined as a process involving relationships between humans, procedures, machines, and the environment. It is crucial to understand these interrelationships that occur when considering the ability of the system to function properly for the mission of the product or system.   In Chapter 3, we have seen the progression from mechanical reliability to attempts at human reliability in the nuclear and other high-risk industries. As stated earlier in Chapter 1, the 1978 publication of *Man-Made Disasters* by the late U.K. sociologist Barry Turner and the 1984 publication of *Normal Accidents,* by the Yale University organizational theorist, Charles Perrow, set the stage for transforming ideas about safety. Turner's book laid the theoretical understanding of the role of organizational factors in accident causation and a revised edition was published in 1997, with Professor Nick Pidgeon.

In the paper, *Limits to Safety,* Pidgeon, examined three recurring aspects of the complex relationship between culture and safety which he believes have been aptly summarized by Professor Lee Clarke, Department of Sociology, Rutgers University: [6]

> Organizational cultures may be organized to enhance imaginations about risk and safety. But they can also insulate organizational members from dissenting points of view. And organizational cultures can perpetuate myths of control and maintain fictions that systems are safe.

Pidgeon concluded his paper with the observation that the first aspect of this complex relationship is the intrinsic cultural blindness to emergent and incubating hazards and institutional design must address ways of circumventing or mitigating that blindness. The second is the realization that discussions about safety culture often fail to fulfill this goal, and presenting an uncritical gloss on the realities of organizational life. While safety and culture hold an intimate relationship, the latter should be invoked only as one part of a wider critique of organizational politics and performance. Both culture and organizational learning are of value for safety practice if viewed from a critical stance. And by bridging the gap between theories of vulnerability and those of resilience, we might come to realize that there are limits to safety—but they may not be quite as narrow as we like to think.

In the 1990s, a number of books concerned with organizational culture appeared. In 1990, the book *Human Error,* by the psychologist, James Reason, modeled the contributory role of managerial weaknesses and human error in catastrophic loss. In 1992, *Organizational Culture and Leadership,* by Edgar H. Schein observed that one major reason for the increased interest in culture is that the concept not only has become relevant to organizational level analysis, but also has aided understanding of what goes on inside an organizations when different subcultures and occupational groups must work with each other. Many problems that were once viewed simply as "communication failures" or "lack of teamwork" are now being understood as a breakdown of intercultural communications. The 1997 publication of another of Reason's books, *Managing The Risks of Organizational Accidents,* provided safety practitioners with the tools to predict where the breaches in risk management defenses might exist. In 1996, The Aviation Foundation and The Institute of Public Policy, George Mason University, released a report that examined the changes that must take place at the FAA so it can regulate an industry that is in constant flux and will be facing significant challenges in the future. The report (which included a case study of the problems with the cancelled Microwave Landing System program) concluded that the primary problem is a culture that does not recognize or serve any client other than itself. "This has fostered a system in which the normal checks and balances do not apply, so there is little accountability in the system. Furthermore, no effective mechanism is in place whereby the FAA can learn from its mistakes and make effective, substantive change."[7] The notion of organization culture was fully aired at a 1997 conference on *Corporate Culture and Transportation Safety,* sponsored by the National Transportation Safety Board. At which, Dr. Reason stated[8] that a "safety culture" is something that emerges gradually from the persistent application of practical measures. For him, Uttal's 1983 definition of organizational culture best describes it nature:

> Shared values (what is important) and beliefs (how things work) that interact with an organization's structures and control systems to produce behavioral norms (the way we do things around here).

In 1994, at the request of the FAA, the National Research Council (NRC) established a Panel on Human Factors In Air Traffic Control Automation to study the human factors aspects of the nation's air traffic control system and future automation issues. The NRC's 1997 publication *Flight To The Future* provides a comprehensive assessment of the role of the air traffic controller in the safety and efficiency equation. The primary focus of the study is the relationship between the human and the automation tools provided to assist in accomplishment of system tasks. The panel adopted a specific interpretation of the term *automation.* It was defined to mean the replacement by machines (usually computers) of tasks previously done by humans. The panel distinguished this term from that of *modernization*, which includes upgrades of air traffic control technology. The impetus for the study grew out of the House of Representatives, Subcommittee on Aviation of the Public Works and Transportation Committee that air traffic control system modernization should not compromise safety and efficiency by marginalizing the role of the human.[9]

Although the *Federal Aviation Regulations (FARs)* state that quantitative assessments of the probabilities of crew error are "not considered feasible." Nonetheless, there is substantial amount of on-going research being undertaken by the UK Civil Aviation Authority (CAA) on a Human Hazard Analysis technique. According to Hazel Courteney, the technique resembles a FMECA or HAZOP analysis based around human (instead of technical) failure modes. Both the FAA and the CAA have formed a harmonization working group to address human factors issues in Type Certification. The Redmill text should be of assistance. The airline industry is also actively involved in human factors research. For example, during the 17th International System Safety Conference, on a Panel "Selected Critical Issues In Aviation Safety Risk Management," Chris Hallman, Flight Training Supervisor, Delta Airlines, presented a model Pilot Risk Management training program designed to improve operational safety margins by focusing on risks and capabilities.

In the previously cited book, *Blind Trust,* Nance made the observation that "before deregulation in 1978, the NTSB, and the FAA were all far behind…in understanding the vital, inextricable role of human factors in the safety equation."[10] Nance stated that "Harvard Professor Ross A. McFarland, generally acknowledged as the "father" of the human-factors field as it relates to aviation, published two landmark books on the subject, *Human Factors In Air Transport Design,* (New York: McGraw-Hill, 1946); *Human Factors In Air Transportation,* (New York: McGraw-Hill, 1953). Professor McFarland taught that for information transfer to take place, any control or other item on an aircraft…that must be read, monitored, manipulated, or otherwise operated by a human being must be designed so that the individual's physical, psychomotor, and psychological capabilities and senses are compatible with it, and so that he or she will be able to interact with the machine as the designers intended.[11]

In 1988 the edited textbook, *Human Factors In Aviation,* by Dr. Earl Wiener, professor of management science and industrial engineering at the University of Miami and Dr. David Nagel of the Aerospace Human Factors Division at NASA' Ames Research Center further advanced knowledge of the field. According to the authors, human factors as a formal independent discipline emerged relatively recently. In the late 1940s in the United Kingdom, an interdisciplinary group of scientists with experience during the war effort formed the Ergonomics Society in 1950 concerned with the human aspects of the working environment. Professor, K.F.H. Murrell derived the word *ergonomics* from the Greek meaning "the science of work." In the United States in 1957, the Human Factors Society was formed. Essentially, the terms ergonomics and human factors are synonymous. At Cambridge University in 1939, the head of the psychological laboratory, Professor, Sir Frederick Bartlett managed a research program that included contributions in the areas of aircrew selection and training, the effects of sleep loss and fatigue, and various aspects of visual perception and display design.   Human factors or ergonomics may be defined as:[12]

> The technology concerned to optimize the relationships between people and their activities by the systematic application of the human sciences, integrated within the framework of system engineering.

Professor Wiener and Dr. Nagel have included the works of a number of aviation-related human factors practitioners. Their textbook also includes an excellent chapter by C.O. Miller on System Safety. In 1990, another representative text on the human factor in aviation, *Flightdeck Performance,* examined what is known about human behavior that is relevant to the primary task of designing aircraft to make them safer to fly and selecting and training pilots to achieve the same goal.[13]   Both texts contain excellent bibliographies that emphasize the practical nature of human factors.   In the 1996 textbook, *Human Factors In Systems Engineering,* the author, Alphonse Chapanis, approached design from the standpoint of how systems should be designed to take into account human considerations. The book opened up with a chilling quotation taken from the remarks made by Charles O. Hopkins, Technical Director of the Human Factors Society Study Group that investigated the Three Mile Island nuclear power plant after the near disaster of March 28, 1979:[14]

> The disregard for human factors in the control rooms was appalling. In some cases, the distribution of displays and controls seemed almost haphazard. ... There were many instances where information was displayed in a manner not usable by the operators, or else was misleading to them. A textbook example of what can go wrong in a man-machine system when people have not been taken into account.

Turning to the operating environment, clearly we see that the "systems" viewpoint has resulted in an omni-disciplinary nature of aviation safety. Consider briefly, the tragic accident of Eastern Air Lines, Flight 375, on October 4, 1960 at Boston's Logan International Airport.   The aircraft had just become airborne when it struck a flock of starlings, and a large number of the birds were ingested into three of its four engines. As a result of this accident, the FAA initiated a research program aimed at improving the tolerance of turbine engines to bird ingestion.[15] At the U.S.

Air Force, in the aftermath of the 1995 AWACS birdstrike tragedy, Wright Laboratory decided to accelerate its search for technologies to prevent serious birdstrikes. The research focuses on the active projection of noise normally associated with aircraft engines to disperse birds from the flightpath. International research is also being conducted to determine *if engine sounds actually convey some biological meaning or warning to birds.* There is some evidence that the noisier engines have lower birdstrike rates. [16]

Safety Engineering Techniques

The System Safety Engineering discipline has embraced traditional reliability engineering tools. The System Safety Society's *Analysis Handbook* lists over 100 analytical tools. The Society has also been at the forefront to seeking a fundamental *shift in thinking* about accidents and their causes. While engineering analysis that focuses on failures and their "failure modes" will continue to be a valuable tool, fatal accidents can and have been caused by a unit functioning as designed without failing to meet its design requirements. The philosophical shift in thinking about safety, the sources of risk and their mitigation can be stated in the following axiom: *When thinking about safety, ask not only what can fail, ask what can go wrong.* This beneficial pessimism is best achieved with the System Safety discipline. It requires the identification and understanding of all known hazards and their associated risks; and mishap risk eliminated or reduced to acceptable levels. The objective of System Safety is *"to achieve acceptable mishap risk through a systematic approach of hazard analysis, risk assessment and risk management.* This emphasis away from a failure-centric perspective has taken on a new urgency because of the increase in software-intensive systems and holds vast implications for our understanding of accidents and safety risk management. System Safety is the assurance that the system in question is safe. For many "Transportation Safety Engineers/Analysts," that assurance begins and ends with analysis of the functions (components) associated with a product or service to be considered. In so doing, there is great reliance on quantitative methodologies such as Fault Trees, Functional Hazard Analyses, and Failure Modes and Effects Analysis, and sophisticated Markov Analysis of devices that exhibit constant failure/reliability rates. In 1962, an analytical tool, Fault Tree Analysis (FTA) was developed by Bell Telephone

Laboratories for the U.S. Air Force.   It is a very detailed analytical technique for determining the various ways in which a particular failure could occur.   There are two basic approaches to FTA: The qualitative approach uses deductive logic to determine the ways in which the undesired "top event" or accident could occur.   The quantitative approach adds reliability or probability of failure data.[17]

In 1973, during the "Golden Decade," William G. Johnson, by developing MORT, The Management Oversight and Risk Tree, broadened the scope of the System Safety discipline by developing standardized monitoring, control of safety, mishap investigation and formal accident report writing for the Atomic Energy Commission. The U.S.Department of Energy has adapted MORT and revisions to the 1,500-item fault tree used for mishap investigation.

The publication of Dev Raheja's textbook, *Product Assurance Technologies,* has sought to bridge the gap between theory and practice.[18] The author has conducted training courses for engineers in making products better, faster, and at lower cost. His book is used at the Graduate School of the University of Maryland in its Master's program in reliability engineering.   For Raheja, reliability engineering is a statistical tool; and statistics are mainly used for measuring, analyzing and estimating various parameters. He defines reliability as the probability that a product will perform its intended function for a given time period under a given set of conditions. The definition assumes the product is in working condition at the outset. Most of the true reliability work is done during the development of a design, well in advance of test data availability for statistical manipulation. Inherent reliability is the highest reliability a product will have. However, reliability can be further degraded in the field because of unanticipated stresses or environments, inadequate procedures, and poor maintenance.   In a highly technical discussion of the so-called "Bathtub Curve" a major analytical tool of Reliability Engineers, Raheja provides a degree of illumination not seen in any of the other books. To understand reliability, it is necessary to understand the concept of the Bathtub Curve. The curve is a plot of hazard rate versus product life. As can be seen in Figure 4:1 The Bathtub Curve, it consists of three regions (Region 1 failures are called infant mortality rates; Region 11, useful design life; and Region 111, wearout). Each of these regions has different hazard rates.

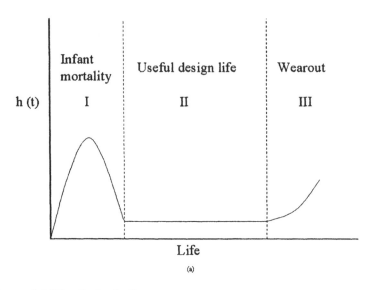

Figure 4:1 The Bathtub Curve

Raheja believes that there is a general misunderstanding about failures in the useful design life region. For a good design, this region has a very low hazard rate and the curve is generally flat (indicating a constant failure rate). The misunderstanding stems from the fact that "many practitioners assume that the region always has a constant hazard rate" *even though the hazard rates for hardware and software are different.* The author also believes that another misunderstanding is the issue of mean time between failure (MTBF), which is used to simplify computations. MTBF is the reciprocal of the hazard rate, as a measure of reliability. Yet, "many misunderstand this term to mean an equivalent of life…when in fact there is no relationship…MTBF is only a function of design and is totally independent of the life or length of the bathtub." For him, this misunderstanding is the reason why failure predictions made in industry rarely correlate with field results. The typical failure rate for mechanical equipment (b); and software (c) versus the life bathtub profile:

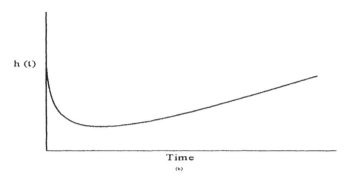

Figure 4:1(b) Mechanical Equipment

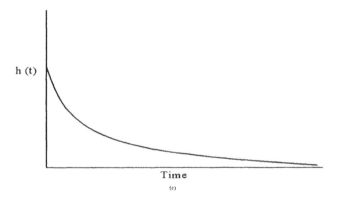

Figure 4:1(c) Software

The Roland and Moriarty textbook also describes applicable management and safety analysis methods. It includes several management decision and evaluation tools from engineering and business that are useful in conducting safety evaluation and risk analysis. The text also helped to clarify a number of terminology issues in the System Safety discipline. Consider for example the following material quoted in Joe Stephenson's *System Safety 2000*,[19] taken from the first two pages of the *(pre-acquisition reform and long since outdated)* 1987 U.S.Air Force's SDP 127-1, *System Safety Handbook for the Acquisition Manager*, stated:

It is difficult to explain the "whys" and "hows"of the System Safety discipline when there is a lack of agreement within the discipline as to just what the task really is. At a meeting of approximately 50 System Safety engineers, each engineer was asked to provide a definition of System Safety. Of those fully qualified and experienced System Safety engineers, at least 30 had distinctly different ideas of what constitutes the system safety task. Very little standardization currently exists between agencies or even between the directives, regulations, and standards that implement the requirement.

The Stephenson text provided a practical guide for planning, managing, and conducting System Safety Programs. Stephenson identified a number of problem areas that need to be addressed before the discipline can provide the safety services to meet the needs of the next century. For him, the lack of standardized terms, basic tools, techniques, safety education and training were problematic. He concluded by advocating that every major engineering program in the country needs at least one System Safety course in its core curriculum. Also, as previously mentioned, the 1997 publication by Nicholas Bahr, *System Safety Engineering and Risk Management: A Practical Approach*, identified the safety educational challenges posed by the lack of formal safety courses in engineering school. Clearly, this startling educational deficiency should be addressed.

In 1996, the National Research Council, in a report titled: *Understanding Risk,* it was noted that risk characterization was often conceived as a summary or translation of the results of technical analysis for the use of a decision-maker. Seen in this light, risk characterization may fail for two reasons: it may portray the scientific and technical information in a way that leads to an unwise decision, or it may provide scientific and technical information in a way that is not useful for the decision maker. Although such failures do occur, an often-overlooked danger to making decisions about risk is "a fundamental misconception about how risk characterization should relate to the overall process of comprehending and dealing with risk."[20] The report defined Risk characterization as "a synthesis and summary of information about a hazard that addresses the needs and interests of decision-makers and of interested and affected parties. Risk characterization is a prelude to decision making and depends on an iterative, analytic-deliberative process."

Analysis and deliberation can be thought of as two complementary approaches to gaining knowledge about the world, forming understandings on the basis of knowledge, and reaching agreement among people. "*Analysis* uses rigorous, replicable methods, evaluated under the agreed protocols of an expert community—such as those of disciplines in the natural, social, or decision sciences, as well as mathematics, logic, and law—to arrive at answers to factual questions. *Deliberation* is any formal or informal process for communication and collective consideration of issues. Participants in deliberation discuss, ponder, exchange observations and views, reflect upon information and judgments concerning matters of mutual interest, and attempt to persuade each other. *Government agencies should start from the presumption that both analysis and deliberation will be needed at each step leading to risk characterization.*[21] (Writer's emphasis). The report provided many thought-provoking ideas about the role of scientific analysis in making sound decisions about risk. It explored the sources of conflicts over risk characterization and how in its choices of assumptions, science is not necessarily neutral. "The assumption of the null hypothesis as used in risk analysis contains an implicit bias because it places a greater burden of proof on those who would restrict than those who would pursue a hazardous activity, presuming these activities are safe until proven otherwise."[22]

The report also provided a discussion on the risk assessment-risk management distinction. *Risk assessment* is the scientific analysis and characterization of adverse effects of environmental hazards. It may include both quantitative and qualitative descriptors, but it often excludes the analysis of the social and economic effects of regulatory decisions. Risk assessment is often presumed to be free of value judgments, with some important exceptions, such as choices about whether and to what extent to include worst-case assumptions in risk assessments. *Risk management* refers to the activities of identifying and evaluating alternative regulatory options and selecting among them. Risk managers are supposed to deal with broad social, economic, ethical, and political issues in choosing decision options supported by the risk assessment.[23]

The Language and Teaching of "Safety"

Three decades ago, Herbert Simon, in a brilliant essay, *The Sciences of the Artificial,* addressed the idea that our thinking about a topic may be deeply rooted in problems of semantics in language processing. Consider that in the collection, reporting metrics and analysis of accident data, the very term "accident" has different meanings. The NTSB defines an *accident* in accordance with international agreements under the United Nation's International Civil Aviation Organization (ICAO) as:

> An occurrence associated with the operation of an aircraft which takes place between the time any person boards the the aircraft with the intention of flight until all such persons have disembarked, and in which any person suffers death or injury as a result of being in or upon the aircraft or by direct contact with the aircraft or anything attached thereto, or in which the aircraft receives substantial damage.

The accident statistics contains data on *non-flight* "accidents" for example, where fatalities occurred while the aircraft was still boarding passengers. Many have argued for a purging of this data to reflect actual flight accidents—which would result in an even lower accident rate. This is an old issue that reappears (predictably after a major airline accident) when aviation experts discuss the "low U.S. accident rate" and inevitable comparisons made to the average 45,000 annual highway deaths as further proof of the safety of aviation. To others, the arguments over accident statistics and *numbers* of people killed appear strange. To begin with, the aviation *numbers* only refer to passengers killed. Crew deaths are not counted. Precisely how many accidents have occurred among previously certified, and therefore presumed to be safe aircraft, is nearly not as important as the fact that the accident investigators found that they were not "Acts of God" and might have been *preventable*. The data, however small, should raise process integrity questions on the artificial dichotomy between certification and safety.

The "Reliability Engineering School" developed an entirely new lexicon appropriate to its methods of analysis. Terms such as "failure," "fault," "failure condition" that may or may not be appropriate when considering accident/incident phenomena of software-intensive systems. Professor Leveson underscores the point that terms in system safety are not used consistently: "The confusion is compounded by the use of the same terms but with different definitions, by engineering, computer science, and natural language."[24] The richly comic character, Humpty Dumpty, readily comes to mind: When I use a word it means just what I choose it to mean—neither more nor less.

We have entered into a *new era* of electro-mechanical systems with an *obsolete safety vocabulary*. Such systems do not necessarily have to "fail" to produce a catastrophe. It is important to communicate material effectively so that students, safety practitioners and safety regulators alike are equipped to ask the right questions when addressing safety risk management issues.   Terms such as "safety," "hazard," "accident," "acceptable level of risk," "certification," have the potential to generate deep distrust.   This has had enormous impact on the ability to conduct *comprehensive* safety assessments on air travel as well as on the level of public discourse and "public acceptance of risks."   Perrow has summed up the societal dilemma posed by the "new breed of shamans, called risk assessors...where body counting replaces social and cultural values and excludes us from participating in decisions about the risks that a few have decided the many cannot do without.... The issue is not risk, but power."[25]

Educational Programs

The System Safety Society has identified a number of "safety degree programs" that are offered by eight colleges (Illinois State; Indiana State; Indiana University of Pa; Marshall Univ.; Millersville Univ., of PA; Murray State; Oregon State and West Virginia Univ.).   The Accreditation Board of Engineering and Technology accredits these.[26] The Society publication also listed the existence of a number of "safety and related degree programs" in 42 states.   The results of a check for colleges that offer safety degree programs revealed that only 6 colleges (Illinois State; Indiana State; Keene State College, NH; Madonna Univ., MI; National Univ., CA; and St. John's Univ., NY) offered Safety Management as a college major.[27]   In 1926, Embry-Riddle Aeronautical University began offering undergraduate programs in aviation.

With regard to short-term safety training programs, in 1952, the University of Southern California (USC) established the first *Aviation Safety Program* at a major research university. The Institute of Safety Systems Management Professional Programs, offers a two-week Certificated System Safety/Risk Management training.   Faculty from three disciplinary areas (engineering, management and psychology) designed the program, part of the USC School of Engineering. A pre-requisite for attendance at the training course is an engineering degree or an undergraduate degree in mathematics. The practical application of probability theory is a major component. Trainees at this intense program delve right into the various quantitative analyses including, Boolean Algebra, binomial, multinomial and hypergeometric distributions, Poisson distribution etc.

The National Surety Training Center at Sandia National Laboratories offers a System Safety Training course for managers and decision makers who have responsibility and accountability for programs that have the potential for catastrophic loss. The course consists of several modules including basic system safety, analysis tools and methodologies, human factors, software safety, root cause analysis and risk communication. The National Aeronautics And Space Administration (NASA) Safety Training Center (NSTC) was established in 1991 by the NASA Headquarters Safety Directorate. Workshops to develop student familiarity with hazard identification hazard analysis, and risk management are conducted. The University of Houston, Clear Lake, approves the program. The program includes concepts in risk management, working with the risk analysis matrix, failure modes and effects analysis, fault tree analysis, event tree analysis and cause-consequence analysis. After the *Challenger* disaster, NASA made a commitment to the use of probabilistic risk analysis. The U.S. Department of Transportation operates the Transportation Safety Institute. In 1998, the FAA arranged a two-year Intergovernmental Personnel Act assignment for Dr. George Donohue, Associate Administrator for Research and Acquisitions, to George Mason University to develop a graduate-level Aviation Transportation System Engineering Certificate Program. Also in1998, The George Washington University implemented a new Aviation Safety and Security Program. Professor Leveson's course: An Introduction to Software System Safety, emphasizes that safety is not a property of software, but a property of the system; and includes topics on software, systems, quality assurance, and safety. Leveson believes that practitioners will find the information necessary to design a safety program, but a cookbook approach is not provided. One set of procedures that applies to all systems and all organizations does not and cannot exist.

Not surprisingly, none of the above safety training courses provide a comprehensive, "big picture" overview or survey of the historical context of U.S. transportation safety. This is to be expected since they were not designed to do that. Neither do they address the discoveries brought about by this literature search and historical classification; nor do they address issues of risk communications. Here again, by their very nature, the majority of two-week safety training programs concentrate on classical analytical techniques. A central concern of this essay is that while necessary, they are not sufficient for safety education.

A Model Aviation Safety Risk Management Process

The 1988 report *Safe Skies,* called on the FAA to establish "a powerful system safety program"[28] for enhancing air safety. The agency then began the process of establishing the elements of a System Safety program that included developing an agency Order and a draft *System Safety Handbook.* Let us briefly examine the FAA's Risk Management and System Safety Process.[29]

On June 26th 1998, the FAA Administrator signed an Agency Order on the subject of Safety Risk Management. The Order established the Agency's safety risk management policy and prescribed procedures for implementing safety risk management as a decisionmaking tool within the FAA. Essentially, the Order calls for use of a formal, disciplined, and documented decisionmaking process to address safety risks in relation to high-consequence decisions impacting the complete product life cycle. The approach to safety risk management is composed of the following steps:

- Plan. A case-specific plan for risk analysis and risk assessment shall be predetermined in adequate detail for appropriate review and agreement by the decisionmaking authority prior to commitment of resources. The plan shall additionally describe criteria for acceptable risk.
- Hazard Identification. The specific safety hazard or list of hazards to be addressed by the safety risk management plan shall be explicitly identified to prevent ambiguity in subsequent analysis and assessment.
- Analysis. Both elements of risk (hazard severity and likelihood of occurrence) shall be characterized. The inability to quantify and/or lack of historical data on a particular hazard does not exclude the hazard from this requirement. If the seriousness of a hazard can be expected to increase over the effective life of the decision, this should be noted. Additionally, both elements should be estimated for each hazard being analyzed, even if historical and/or quantitative data is not available.

- Assessment. The impact of the risk elements in the analysis shall be compared to acceptability criteria for decisionmaking.
- Decision. The risk management decision shall consider the risk assessment results conducted in accordance with the assessment. Risk assessment results may be used to compare alternative options.
- Feedback. Feedback on performance is a critical element.

FAA's System Safety Process

The FAA's system safety process is a logical, engineering approach for obtaining system safety objectives, as shown in Figure 4-2. This closed loop process is tailorable and can be applied at any point in the system life cycle, but the greatest advantages are achieved when applied early in the acquisition phase.

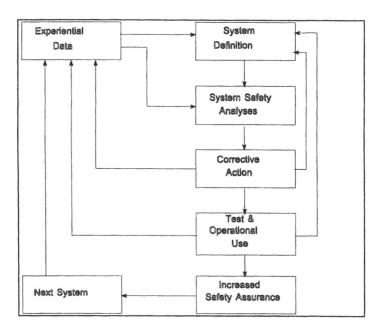

Figure 4:2 FAA System Safety Process

This process is normally repeated as the system evolves or changes as problem areas are identified.   As can be seen from the following descriptions, a number of key variables.

Experiential Data.   Experiential data represent corporate memory/lessons learned or knowledge gained from operation of previous similar systems. Experiential data can be useful as inputs to the preliminary hazard analysis. Previous action taken to correct design features that have resulted in accidental damage, loss, injury, or death. This corrective action includes design changes, production/operational retrofits, and operating/maintenance procedure changes.

System Definition.   The first step in the process is to clearly define the system under consideration. System definitions must also include major system interfaces such as system operating condition, environmental situation, and the human role in system operation.   The object is to set limits for the following steps in the process and reduce complex systems to manageable parts.

System Safety Analyses.   The heart of the system safety process is a comprehensive, methodical analysis of the system and its elements.   The search for possible system hazards is the state of before-the-fact *accident prevention.*

The analyses *must* consider every undesired event that might occur to the system and either the conditions that produce the undesired event or the consequences from it. This requires a methodical or systematic approach for the analyses. A thorough analysis should identify possible hazards, classify the hazard for severity, determine the probability of occurrence of the hazard, and suggest possible corrective action.

The effectiveness of corrective action is also a part of the analysis. It is essential to maintain a closed loop hazard identification and tracking system so that all identified hazards are followed through the corrective action or program decision on identified hazards. This type of documentation, while not part of the analyses, is the administrative component of the System Safety Program.

Corrective Action to Eliminate/Control Hazard.   The three steps immediately above identify and assess the hazard risks but do not eliminate or control them.   It is ultimately the responsibility of the program manager to see that corrective action is taken. This is perhaps the most crucial step in the entire process.   Any action taken will modify or change some element of the system.   The modification may involve hardware, software, procedures and/or training. If the system is modified, the initial definition of the system elements and assumptions should be revised. The process is then repeated, as required, for potential additional hazards introduced by system modifications to ensure that actions taken to correct one hazard do not induce more hazards elsewhere in the system.

Test and Operational Use. The previous steps identified hazards through analysis and controlled or eliminated them (within program limitations). In an imperfect world, all hazards may not be identified in this process. Others can and will be identified when the system is exercised through test and operational use. Any occurrence of an accident or incident is examined critically to determine causes and evaluate effects. The causes and effects could range from something already predicted as possible or even probable under certain conditions, to something entirely new and surprising.

The results of mishap analysis may reveal other deficiencies in the system design or procedures and serve to direct corrective action back to the system safety process. In this way, maximum use is made of the mishap experience without having to go back and continually rediscover new truths. Most, if not all, development programs for complex systems include testing to verify performance and the demonstration of system capabilities. They are conducted to assure the user that this system performs as required. Tests and demonstrations normally performed on a system or its components are also planned and conducted to reveal any safety inadequacies.

Concluding Remarks

While engineering analysis that focuses on failures and their "failure modes" will continue to be a valuable tool, fatal accidents can and have been caused by a unit functioning as designed without failing to meet its design requirements. The philosophical shift in thinking about safety, the sources of risk and their mitigation can be stated in the following axiom: *When thinking about safety, ask not only what can fail, ask what can go wrong.'* The following provide some practical lessons.

- *The FAA System Safety Handbook* has tailored the Mil Standard 882 process to aviation. At the time of this writing, it is being revised. The

Roland and Moriarty text, *System Safety Engineering And Management,* defined System Safety as the application of special technical and managerial skills to the systematic, *forward-looking* identification and control of hazards throughout the life cycle of a project.

The award winning paper: "Beauty And The Beast – Use And Abuse Of The Fault Tree As A Tool" by a Senior System Safety Engineer, described proper application and misapplications of the fault tree as a tool when evaluating complex systems. Safety practitioners should become familiar with Hammer's earlier works, *Handbook of System and Product Safety,* and *Occupational Safety Management And Engineering,* for practical applications. Hammer observed that the most effective means to avoid accidents during system operation is by eliminating or reducing hazards and dangers during design and development. Hammer's work served as an inspiration for many authors. Nancy Leveson's, text, *Safeware: System Safety and Computers,* is a major contribution to our understanding of the role of software in high-risk systems.

Dev Raheja's textbook, *Product Assurance Technologies,* provides an excellent discussion of the misunderstandings surrounding the differing failure rates for hardware and software and explanations of analytical techniques. *System Safety: HAZOP and Software HAZOP,* provides safety practitioners with a detailed discussion and practical examples of the hazard and operability (HAZOP) technique to the identification and analysis of hazards in software-based systems.

- Steven M. Casey's *Set Phasers On Stun And Other True Tales of Design, Technology, and Human Error,* presented twenty factual accidents to serve as vehicles for safety thinking. Readers are required to draw their own conclusions. Turner's *Man-Made Disasters,* which preceded Charles Perrow's *Normal Accidents,* are both strongly recommended for perspectives on how organizational processes have contributed to accidents.

- James Reason's *Managing The Risks of Organizational Accidents,* provided safety practitioners with the tools to predict where the breaches in risk management defenses might exist. The text, *Beyond Aviation Human Factors,* by Daniel E. Maurino, James Reason, Neil Johnson, and Rob B. Lee provides new thinking about the role of an information-based "organizational model" for accident prevention.

Notes

[1]Willie Hammer, *Handbook of System and Product Safety*, (Englewood Cliffs, NJ: Prentice-Hall, Inc., 1972).

[2]IBID., p. 21.

[3]This following paragraphs are adapted from Harold E. Roland & Brian Moriarty, *System Safety Engineering And Management*. (New York: John Wiley & Sons, Inc., 1990). 2nd ed., pp. 8-11.

[4]Leveson, op. cit., pp. IX.

[5]Felix Redmill; Morris Chudleigh, and James Catmur, *System Safety: HAZOP and Software HAZOP*, (West Sussex, England: John Wiley & Sons Ltd., 1999), pp.155.

[6]Nick Pidgeon, *The Limits To Safety? Culture, Politics, Learning and Man-Made Disasters*, Op.Cit.

[7]*Why Can't The Federal Aviation Administration Learn? Creating A Learning Culture At the FAA*, The Aviation Foundation and The Institute of Public Policy, Falls Church, VA and George Mason University, Fairfax, VA. PP. 7. July 10,1996.

[8]James Reason, "Corporate Culture and Safety," Conference Proceedings *Corporate Culture and Transportation Safety*, sponsored by the National Transportation Safety Board, April 24-25, 1997.

[9]National Research Council, *Flight To the Future*, (Washington, DC: The National Academy Press, 1997).

[10]John J. Nance, Op. Cit., p. 337.

[11]IBID., p. 75.

[12]Earl Wiener and David Nagel, Eds., *Human Factors In Aviation*, (San Diego,CA: Academic Press, Inc., 1988), p.9.

[13]David O'Hare and Stanley Roscoe, *Flightdeck Performance*, (Ames, IA: Iowa State University Press, 1990).

[14]Alphonse Chapanis, *Human Factors In Systems Engineering*, (New York: John Wiley & Sons, 1996), p.1.

[15]*Aviation Disasters*, 2nd ed. pp. 37-39.

[16]Jeffrey J. Short et. al. "Recent Research Into Reducing Birdstrike Hazards." U.S. Department of Energy. See also John Thorpe: "Bird Strikes To Airliner Turbine Engines." U.K. Civil Aviation Authority. Papers presented at the London Meeting: Bird Strike Committee Europe, May 13-17,1996.

[17]Joe Stephenson, *System Safety 2000*, (New York: Van Nostrand Reinhold, 1991), p.167.

[18]Dev Raheja, *Product Assurance Technologies*, (Laurel, MD: Technology Management Inc., 1995), pp. 39-41.

[19]Quoted in Joe Stephenson, *System Safety 2000*, (New York: Van Nostrand Reinhold, 1991), pp.43-44.

[20]*Understanding Risk: Informing Decisions In A Democratic Society*, (Washington, DC: The National Academy Press, 1996), pp.1.

[21]IBID., pp. 3-4.

[22]IBID., p. 25.

[23]IBID., p.33.

[24]Nancy Leveson, *Safeware: System Safety and Computers.* (Reading, MA: Addison-Wesley Publishing Co. 1995), p. 171.

[25]Perrow, Op. Cit., p.12.

[26] *System Safety Analysis Handbook,* 2nd edition, July 1997.

[27] Barron's *Profiles of American Colleges*, 1995 Ed.

[28] *Safe Skies,* Op.Cit., p. 4.

[29] Adapted from: *Draft FAA System Safety Handbook,* Draft Chapter 2 (Developed by Steven D. Smith, FAA Office of System Safety Engineering & Analysis).

5 Conclusion

"If we are to retain any command at all over our own future, the ablest people we have in every field...have to come out of the trenches of their own specialty and look at the whole battlefield."

John Gardner

There is no doubt that this literature guide has shown an upsurge of interest in air transportation safety with an ever-broadening range of critical thinking and policy implications. The material has been presented in a conceptual framework for the benefit of safety practitioners, students, as well as the general public who wish to further explore the area. Three major "Schools" of thought and their philosophical underpinnings have been identified and form the basis of the chapters. In presenting the transportation safety literature, what has clearly stood out is its ever expanding, omni-disciplinary nature that has given, and continues to give rise to, alternative thinking and new ways of measuring "safety" and in addressing what can be done about preventing accidents.

A fundamental philosophical underpinning of the "Reliability Engineering School" has been the ability to describe the probability of future events, such as equipment failure, with numbers—as in one chance in a million—made it possible to accurately compare the frequency and severity of various risks. The essence of risk analysis is the application of rigorous mathematical analysis to enable risk managers and regulators alike to decide what to do under periods of uncertainty, in full knowledge that we can never arrive at a state of perfect "safety." The "System Safety Engineering School," while incorporating the tenets of failure-centric thinking, posited a major philosophical change away from reliability engineering. The core objective of System Safety engineering is the task to first identify hazards and, through design, try to *prevent* them from causing harm or try to reduce the risks to acceptable levels. In so doing, System Safety engineering developed a system of accountability that would hold an enormous presence in product liability lawsuits. The System Safety discipline emphasized that in the era of software-intensive systems, fatal accidents can and have been caused by a unit functioning as designed without failing to meet its design requirements. Thus, when thinking about safety, ask not only what can fail, ask what can go wrong.[*]

The emerging literature on modern organization theory, human factors research and corporate safety culture suggests that administrative and organizational failures are elements that can also go wrong. Thus, to meet the challenge to produce truly comprehensive safety analyses that measurably improve our ability to identify safety risks stemming from all possible sources, the focus of our hazard identification efforts should include hardware, software, as well as administrative and organizational dysfunction. New knowledge from newspaper accounts and the Internet about bizarre air transportation incidents under investigation have heightened our awareness of system capacity and safety issues. Consider the reports on how a spilled coffee cup caused a near-collision between arriving and departing planes; and the "inadvertently" tripped electric switch that shut down power to communications in the control center and caused controllers to lose radio contact with about 100 planes for 10 minutes when the backup radio system failed. Competent leadership in the safety profession goes far beyond the narrow confines of technical specialties or the routine block-completion exercises on a hazard/failure analysis safety management checklist, to include a far broader requirement to fuse the multiple streams and patterns in safety thinking that this literature survey has discovered. Dr. James Reason's practical admonition that "it only takes one organizational accident to put an end to all worries about the bottom-line" clearly addresses the economics of safety. Also, we have seen that many individuals populate the safety discipline and through their actions, as documented in this essay, have made a difference: from Lorenzo Coffin, to Garrett A. Morgan, Ralph Nader, Lois Gibbs, John King, Mary Schiavo, Janette Fennell, Richard M. Cooper and Judith C. Areen, who have all helped redefine corporate and governmental regulatory accountability for safety.

---

[*] The respected U.K.-based aviation safety consultant, David Gleave, Aviation Hazard Analysis Inc., has stated that he is guided by the following quotation from Kipling: "I keep six honest serving men (They taught me all I knew); Their names are What and Why and When and How and Where and Who. I send them over land and sea, I send them east and west; but after they have worked for me, I give them all a rest.

Challenges of an Omni-disciplinary Profession

The 1997 textbook, *Commercial Aviation Safety,* by Alexander T. Wells, a professor at Broward Community College, provided a study of how major elements of aviation safety: the man, machine, medium, mission, and management factors—the 5-M Factors, prevent accidents and incidents. In cooperation with the University Aviation Association, the book contains an extensive chapter on the subject of airline management in aviation safety. Professor Wells observed that until the early 1990s…only about 50 percent of U.S. airlines had identifiable safety departments. The position of Safety Manager is much like an ombudsman and is critical to effective risk management.[1]

Consider briefly that Part 119.65 of the *Federal Aviation Regulations, Management Personnel Required For Operations Conducted Under Part 121,* stipulate that each certificate holder must have sufficient qualified management personnel to ensure the highest degree of safety in its operations. The certificate holder must have qualified personnel serving full-time in the following or equivalent positions: Director of Safety; Director of Operations; Chief Pilot; Dispatcher of Maintenance; and Chief Inspector. Curiously, with the exception of Director of Safety, the regulations specify in fairly minute detail the qualifications for all the other positions. It is as though "Safety Director" is an indefinable abstraction, validating Vern Grose's observation that since no one really knows what to do about "safety," safety gets assigned to the four winds—in the hope that since everyone is responsible, no one can be blamed for accidents. Is it that the position of "Airline Safety Director" is such an amorphous amalgamation of duties that it can be done by anyone, hence there is no need for its description? Or that because of its apparently universal understanding, the position of "Safety Director" is not worthy of further definition? The NTSB does not agree and has specifically asked the FAA to adopt regulatory action to specify the "form, structure, and function of an air carrier safety department." Finally, on November 30th, 1999, some five years after Secretary of Transportation Federico Pena's Aviation Safety Summit Meeting, the FAA published "guidance material" for the position of "Aviation Safety Director."

In 1979, the noted Public Administrator, Harlan Cleveland, former dean of the Maxwell School, Syracuse University, and past-president of The American Society for Public Administration, made an observation about administration which has a great deal of relevance to the safety discipline. Dean Cleveland noted that the art of administration is above all the executive's willingness to hold contradictory propositions comfortably in a mind that relishes complexity. He described the central paradox of large-scale administration (as stated by Isaiah Berlin in one of his *Conversations* with Henry Brandon) in five incandescent sentences:[2]

> As knowledge (becomes) more and more specialized, the fewer are the persons who know enough ... about everything to become wholly in charge ... One of the paradoxical consequences is therefore the dependence of a large number of human beings upon a collection of ill-coordinated experts, each one of whom sooner or later becomes oppressed and irritated by being unable to step out of his box and survey the relationship of his particular activity to the whole. The experts cannot know enough. The coordinators always did move in the dark, but now they are aware of it. And the more honest and intelligent ones are rightly frightened by the fact that their responsibility increases in direct ratio to their ignorance of an ever-expanding field.

Dean Cleveland went on to note that "none of us is trained for the scary profession of managing more while knowing less. No university in the world offers a Ph.D. in Getting It All Together." Many interdisciplinary courses that were "team-taught" meant that three or four professors shared the task of teaching the same group of students. The result was that each teacher teaches his or her own discipline. It is the students who are expected to be interdisciplinary. In short, everyone knew that "the only true interdisciplinary instrument is not a committee of experts, but the synoptic view from a single integrative mind." To be a get-it-all-together person is not a profession; it is an attitude toward all professions, a propensity to interest oneself especially in the interconnections among the traditional jurisdictions into which we have divided the life of the mind, a willingness to view every problem in a global perspective. And the presumption to feel personally responsible for the whole outcome of which any individual's efforts can only be a small part.

In the award winning conference paper[3] on fault tree analysis, R. Allen Long, a Senior System Safety Engineer lamented the fact that most disciplines tend to see only their own tiny areas of expertise. Few engineers and other highly technical personnel can divorce themselves from their specific areas to look at the larger picture.... The interactions and interfaces are dealt with as necessary inconveniences to getting their particular part connected to the overall system.... This has led to a plethora of FTAs ...simply to satisfy a contractual or management requirement. The term "integration" has been reduced to a buzzword:

> Engineering groups are largely compartmentalized only "interfacing" with each other after they have completed their respective part of a project. Making the pieces fit together is an afterthought—usually addressed too late in the design process to provide a truly integrated product or process. Each engineering discipline basically bolts on its piece of hardware to the project. It is this "cross-strapping" of systems where the fault tree can pay for your effort. This is the most likely place where problems have not been adequately addressed nor discovered by the various designers. This is where other analysis methodologies and engineering disciplines fail and where the system safety engineer can convince other engineering groups that we are a legitimate "engineering" discipline.

Those qualities of mind outlined by Dean Cleveland that facilitate the viewing of problems in a global perspective, are ideally desirable traits for a corporate-level Airline Safety Director. The key is the *attitude* towards the safety profession with experience, interest and communications capability among the interconnected positions of Director of Operations; Chief Pilot; Dispatcher of Maintenance; and Chief Inspector. Drawing from this *literature guide*, and "best practice," it is evident that an ideal set of job requirements would include a satisfactory combination of technical and non-technical skills, experience, training, education and knowledge in the following areas:

- Flight safety
- Reliability calculations / statistics / information management
- Risk management / tort law
- Advising and highlighting unsafe practices / software safety
- System Safety Engineering principles
- Regulatory philosophy of transportation safety
- Economics of safety

Three decades ago, Herbert Simon's essay, *The Sciences Of The Artificial,* argued that engineers are not the only professional designers. "Everyone designs who devises courses of action aimed at changing existing situations into preferred ones.

The intellectual activity that produces material artifacts is no different fundamentally from the one that prescribes remedies for a sick patient or the one that devises a new sales plan for a company or a social welfare policy for a state.   Design so constructed, is the core of all professional training; it is the principal mark that distinguishes the professions from the sciences. Schools of engineering, as well as schools of architecture, business, education, law, and medicine, are all centrally concerned with the process of design."[4] Simon examined the 1930s reform movement at the Massachusetts Institute of Technology in the engineering curriculum toward natural sciences and urged in his essay for the inclusion of the fundamentals of design.

For a 1990s perspective on the state of the engineering design profession, Dr. George Hazelrigg, Project Director, Design and Integration Engineering Program, National Science Foundation, stated in a telephone interview that the number of Doctoral dissertations in engineering design was "weak and only a handful of colleges (Georgia Tech, Ohio State and Michigan) produced anything of substance."   According to him, prior to the Second World War, experienced engineers taught engineering. The science-driven successes of the war resulted in the movement of engineering into the applied sciences essentially at the expense of design. Namely, "we ceased development of a theory of design and the teaching of synthesis. There is nothing physical about design. Design is a thought best carried out in a logical, mathematical framework." Dr. Hazelrigg has authored the textbook, *Systems Engineering: An Approach To Information-Based Design.* (Prentice-Hall, NJ, 1996). It presents a logical approach to engineering design.

The historical rigidity of our major engineering schools in not addressing "safety", as part of the curriculum needs to be re-assessed. Educational assessments of key professions are nothing new. For example, in the wake of the Watergate scandal, all accredited law schools instituted a requirement for a course on professional responsibility.

At MIT, there has been extremely encouraging news for "safety" since Professor Nancy Leveson introduced the discipline of System Safety in the engineering program. It has been noted earlier that at the end of the 20th century some 80 percent of the engineers who are charged with and operating safe systems and are often called upon to testify in courts of law to testify about design failures had never taken a safety course in college nor attended a safety conference. Notwithstanding the tenets of the "Reliability Engineering School," this is an appalling discovery. And stands as testimony for national level attention to the "safety education" component of engineering students. Accordingly the President's Science and Technology Advisor and the National Academy of Sciences, in cooperation with the engineering professional societies, should investigate this unique issue. The earlier referenced Academy's role in defining risk and human factors in air traffic automation and modernization has been highly influential and a similar effort in "safety education" is much needed. Safety is inherently omni-disciplinary. There is need for a mental framework that enables engineering designers, safety managers and practitioners to think beyond vocationalism. As they design and manage large socio-technical systems and communicate the inherent risks to the public, they are guided by the awareness that absolute safety does not exist. All human activity involves risk and the purpose of a safety risk management program is to reduce the likelihood of mishaps to a societally acceptable level:

- If we examine history, an important generalization, which might be called the "*Titanic* effect," can be discerned: The magnitude of disasters decreases to the extent that people believe that they are possible, and plan to prevent them, or to minimize their effects. (Watt).

- One of the main problems in resolving intercultural issues is that we take culture so much for granted and put so much value on our assumptions. ...We tend not to examine assumptions once we have made them ...and we tend not to discuss them...If we are forced to discuss them, we tend not to examine them but to defend them because we have emotionally invested in them. (Schein)

- Organizational cultures may be organized to enhance imaginations about risk and safety. But they can also insulate organizational members from dissenting points of view. And organizational cultures can perpetuate myths of control and maintain fictions that systems are safe. (Clarke).

- Safety is more than the absence of accidents. Safety is the goal of transforming the levels of risk that inheres in all human activity. (McIntyre).

Notes

[1]Alexander T. Wells, *Commercial Aviation Safety,* (New York: McGraw-Hill, 1997), PPS. 216-223.

[2]Harlan Cleveland, "The Get-It-All-Together Profession," *Public Administration Review, Vol.39,* July/August 1979, PP. 306.

[3]R. Allen Long, "Beauty And The Beast – Use And Abuse Of The Fault Tree As A Tool." 17th International System Safety Conference *Proceedings,* August 15-21 1999. PPS. 117-127.

[4]Herbert Simon, *Sciences of the Artificial,* (Cambridge, MA: The MIT Press, 1969), PPS. 55 – 58 *passim.*

Bibliography

Amstadter, Bertram L. *Reliability Mathematics*, (New York: McGraw-Hill Inc., 1971).

Auflick, J.L.; Hahn, H.A.; Morinski, J.A.; "HuRa! – A Prototype Expert System for Quantifying Human Errors," *Probabilistic Safety Assessment And Management (PSAM 4), Conference Proceedings*, Vol. 1.

Bahr, Nicholas J. *System Safety Engineering and Risk Assessment: A Practical Approach*, (Washington, DC: Taylor & Francis, 1997).

Barron's *Profiles of American Colleges*, 1995 Ed.

Bernstein, Peter L., *Against The Gods*, (New York: John Wiley & Sons, Inc. 1996).

British Airways Safety Information System (BASIS). January, 1988.

Captain "X" and Dodson, Reynolds. *Unfriendly Skies*, (New York: Doubleday, 1989).

Casey, Steven M. *Set Phasers On Stun: And Other True Tales of Design, Technology and Human Error*, (Santa Barbara, CA: Agean Publishing Co., 1998), 2nd ed.

Chapanis, Alphonse. *Human Factors In Systems Engineering*, (New York: John Wiley & Sons, 1996).

Clarke, Lee. Department of Sociology, Rutgers University: "The Disqualification Heuristic: When Do Organizations Misperceive Risk? *Sandia Report*. Proceedings of the High Consequence Operations Safety Symposium. July 12-14, 1994. Organizational Strategy and Management Session.

Cleveland, Harlan. "The Get-It-All-Together Profession," *Public Administration Review*, Vol.39, July/August 1979, PP. 306.

Court of Appeals of California.119 Cal. App. 3rd 757: 1981 Cal.App. LEXIS 1859.

Dawkins, S.K., Kelly, T.P., McDermid, J.A. Murdoch J., Pumfrey, D.J. "Issues In The Conduct of PSSA." 17th International System Safety Conference. *Proceedings*, p. 77.

Eddy, Paul, Potter, Elaine, and Page, Bruce. *Destination Disaster*, (New York: Quadrangle/The New York Times Book Co., 1976).

FAA Aviation Forecasts, Fiscal Years 1995-2006.

FAA: Global Analysis Information Network (GAIN). May 1996.

FAA: Land and Hold Short Operations Risk Assessment. September 1999.

FAA System Safety Handbook. (Draft 1995).

FAA Safety Risk Management. Order 8040.4.

Federal Aviation Regulations, (FARs) Part 119.65.

Federal Supplement, Volume 856, PP.727; 1994 U.S. District. LEXIS 8873.

Fischoff, Baruch, Risk Perception and Communication Unplugged: Twenty Years of Process. *Risk Analysis*, Vol.15. No.2, 1995, pp.137-143.

Gero, David. *Aviation Disasters*, (London: Butler & Tanner Ltd., 1996), 2nd ed.

Godson, John. *Unsafe at Any Height*, (New York: Simon and Schuster, 1970).

Graham, John D., Green, Laura C. and. Roberts, Marc J. *In Search of Safety*, (Cambridge, MA: Harvard University Press, 1988).

Grimaldi, John V. and Simonds, Rollin H. *Safety Management*. (Homewood, IL: Richard D. Irwin, Inc., 1975) 3rd Ed.

Green, Roger G., Muir, Helen, James, Melanie, and Gradwell, David. *Human Factors for Pilots*. (Hants, England: Avebury Technical, 1991).

132

Grose, Vernon L. *Managing Risk,* (Arlington, VA: Omega Systems Group, 1987).

Hammer, Willie. *Handbook of System and Product Safety,* (Englewood Cliffs, NJ: Prentice-Hall, Inc.1972).

Hammer, Willie. *Occupational Safety Management And Engineering,* (Englewood Cliffs, NJ: Prentice-Hall, Inc.1981), 2nd Ed.

Hawkins, Frank L. *Human Factors In Flight* (Brookfield, VT: Ashgate Publishing Company, 1987).

Hazelrigg, George. *Systems Engineering: An Approach To information-Based Design.* (Englewood Cliffs, NJ: Prentice-Hall, Inc. 1996).

Heinrich, H.W., Petersen, Dan, Roos, Nestor. *Industrial Accident Prevention,* (New York: McGraw-Hill, 1980) 5th ed.

Henley, E.J. and Kumamoto, H. *Reliability Engineering and Risk Assessment,* (Englewood Cliffs, NJ: Prentice-Hall, Inc. 1981).

Henley, E.J. and Kumamoto, H. *Probabilistic Risk Assessment,* (New York: IEEE Press, 1991).

Holbrook, Stewart H. *The Story of American Railroads,* (New York: Crown Publishers, 1947), "The Airbrake Fanatic."

Holbrook, Stewart H. *Lost Men of American History,* (New York: The Macmillan Company, 1947), "Discontent: The Mother of Progress."

Henderson, Keith M. *Emerging Synthesis in American Public Administration.* (Asia Publishing House, 1966).

International System Safety Conference, *Proceedings,* August 15-21, 1999.

Jones, Richard B. *Risk-Based Management.* (Houston, TX: Gulf Publishing Company, 1995).

Leveson, Nancy. *Safeware: System Safety and Computers,* (Reading, MA: Addison-Wesley Publishing Co. 1995).

Lloyd, E. and Tye, W. *Systematic Safety,* (London, U.K.: Civil Aviation Authority, 1982).

Logan, Rayford. W. and Winston, Michael R., *Dictionary of American Negro Biography,* (New York: Norton & Company, 1982), p.453.

Long, R. Allen. Senior System Safety Engineer, Hernandez Engineering, Inc. Al. "Beauty And The Beast – Use And Abuse Of The Fault Tree As A Tool," 17th International System Safety Conference *Proceedings*, pp. 117- 127.

Lowell, Captain Vernon W. *Airline Safety Is A Myth,* (Bartholomew House, 1967).

Maurino, Daniel E., Reason, James, Johnson, Neil, and Lee, Rob B. *Beyond Aviation Human Factors,* (Brookfield, VT: Ashgate Publishing Co., 1999).

McCall, Brenda. *Safety First at Last.* (New York: Vantage Press, 1975).

McFarland, Ross A. *Human Factors In Air Transport Design,* (New York: McGraw-Hill, 1946).

McFarland, Ross A. *Human Factors In Air Transportation,* (New York: McGraw-Hill, 1953).

Modarres, M., *Reliability And Risk Analysis,* (New York: Marcel Dekker, Inc., 1993.

Mosleh, A. and Bari, R.A. (Eds.) *Probabilistic Safety Assessment and Management.* PSAM 4. Proceedings of PSAM 4 Conference, 13 –18 Sept. 1998, New York.

Nader, Ralph *Unsafe At Any Speed*, (New York: Grossman Publishers, Inc., 1965).

Nader, Ralph / Smith, Wesley J., *Collision Course· The Truth About Airline Safety,* (TAB Books, McGraw-Hill, Inc., 1994).

National Research Council, *Science And Judgment In Risk Assessment,* (Washington, DC: National Academy Press, 1994), Prepublication Copy.

National Research Council, *Understanding Risk: Informing Decisions In A Democratic Society*, (Washington, DC: The National Academy Press, 1996.

National Research Council, *Flight To The Future Human Factors In Air Traffic Control.* (Washington, DC: National Academy Press, 1997).

National Transportation Safety Board, *Conference Proceedings, Corporate Culture and Transportation Safety.* April 24-25, 1997, Washington, DC.

Nance, John J. *Blind Trust*, (New York: William Morrow & Company, 1986).

Neumann, Peter. *Computer-Related Risks*, (Reading, MA: Addison-Wesley Publishing Co. 1995).

NHTSA Report, HS-028 688. *Notre Dame Lawyer*, Notre Dame University: School of Law. 1979/06, PPS. 911-924, "Corporate Homicide: A New Assault on Corporate Decision-Making."

NHTSA Report, HS-030 349. *University of Detroit Journal of Urban Law*, Detroit University, School of Law, 1979. 56 (4) PPS.30. "Beyond Products Liability: The Legal, Social, And Ethical Problems Facing The Automobile Industry In Producing Safe Products."

O'Hare, David. and Roscoe, Stanley. *Flightdeck Performance*, (Ames, IA: Iowa State University Press, 1990).

Pate-Cornell, Elisabeth. Department of Industrial Engineering and Engineering Management, Stanford University. "Priorities in Risk Management: Human and Organizational Factors as External Events and a Maritime Illustration," (PSAM 4) *Conference Proceedings* Vol. 4, pp. 2675.

Perrow, Charles. *Normal Accidents*, New York: Basic Books, Inc., Publishers, 1984).

Petroski, Henry. *To Engineer Is Human.* (New York: 1st Vintage Books Edition, 1992).

Pidgeon, Nick. "Systems, Organizational Learning, and Man-Made Disasters," *Probabilistic Safety Assessment And Management (PSAM 4), Conference Proceedings*, Vol. 4.

Pidgeon, Nick, *"The Limits To Safety? Culture, Politics, Learning and Man-Made Disasters,"* Draft paper prepared for, Special Issue of *The Journal of Contingencies and Crisis Management*, Vol. 5, Number 1, March, 1997).

Powell, Douglas and Leiss, William, *Mad Cows And Mother's Milk*, (Montreal & Kingston, Canada: McGill-Queen's University Press, 1997).

Raheja, Dev. *Product Assurance Technologies*, (Laurel, MD: Technology Management Inc., 1995).

Reason, James. *Human Error*, (Cambridge: Cambridge University Press, 1990).

Reason, James. *Managing The Risks of Organizational Accidents*, (Brookfield, VT: Ashgate Publishing Co., 1997).

Redmill, Felix; Chudleigh, Morris and Catmur, James, *System Safety: HAZOP and Software HAZOP*, (West Sussex, England: John Wiley & Sons Ltd., 1999).

Roberts, N.H., Vesely W.E., Haasl, D.F., and Goldberg, F.F. *Fault Tree Handbook*, NUREG-0492, (U.S.Nuclear Regulatory Commission, Washington, DC, 1981).

Roland, Harold E. & Moriarty Brian. *System Safety Engineering And Management*, (New York: John Wiley & Sons, Inc., 1990). 2nd ed.

Sagan, Scott D., *The Limits of Safety: Organizations, Accidents and Nuclear Weapons*, (Princeton, NJ: Princeton University Press, 1993).

Schein, Edgar H., *Organizational Culture and Leadership*. (San Francisco, CA: Jossey-Bass Inc., 1992). 2nd Ed.

Schiavo, Mary. *Flying Blind, Flying Safe*, (New York: Avon Books, 1997).

Shooman, Martin L. *Probabilistic Reliability: An Engineering Approach.* (New York: McGraw-Hill, Inc., 1968).

Short, Jeffrey J. et. al. "Recent Research Into Reducing Birdstrike Hazards." U.S. Department of Energy. *See also* John Thorpe: "Bird Strikes To Airliner Turbine Engines." U.K. Civil Aviation Authority. Papers presented at the London Meeting: *Bird Strike Committee Europe,* May 13-17,1996.

Simon, Herbert. *Sciences of the Artificial,* (Cambridge, MA: The MIT Press, 1969).

Sprague, Elmer. & Taylor, Paul W. *Knowledge and Value,* (New York: Harcourt, Brace & World, Inc., 1959).

Starling, Grover. *The Politics and Economics of Public Policy,* (Homewood, IL: The Dorsey Press, 1979),

Stephenson, Joe. *System Safety 2000,* (New York: Van Nostrand Reinhold, 1991).

System Safety Society, *System Safety Analysis Handbook, 2nd* Ed. July 1997.

Tarrants, William E. *The Measurement of Safety Performance,* (New York: Garland Publishing, Inc. 1980).

Taylor, Captain Laurie. *Air Travel: How Safe Is It?,* (Oxford: BSP Professional Books, 1988).

Tenner, Edward. *Why Things Bite Back.* (New York: Knopf Publishing, 1995).

The Aviation Foundation and The Institute of Public Policy, Falls Church, VA and George Mason University, Fairfax, VA. *Why Can't The Federal Aviation Administration Learn? Creating A Learning Culture at the FAA.* PP. 7. July 10,1996.

The Guide to American Law, (New York: West Publishing Company, 1984), Vol.10.

The Rand Corporation, *Safety In The Skies.* December, 1999.

The Royal Society, *Risk: Analysis, Perception and Management.* (London: The Royal Society, 1992).

The Royal Society, *Science, Policy and Risk.* (London: The Royal Society, 1997).

Transportation Hearings 1998. Subcommittee of the Committee on Appropriations, House of representatives, 105th Congress, 2nd Session, Part 6, p.231.

Transportation Statistics Annual Report, 1997.

Trivedi, Kishor S. *Probability and Statistics with Reliability, Queuing, and Computer Science Applications,* (New Delhi: Prentice-Hall of India, Private Limited, 1997).

Turner, Barry. *Man-Made Disasters,* (London: Wykeham Publications, 1978).

Turner, Barry and Pidgeon, Nick *Man-Made Disasters,* (Oxford: Butterworth-Heinemann, 1997).

U.S. Congress, Office of Technology Assessment, *Safe Skies for Tomorrow: Aviation Safety In A Competitive Environment,* OTA-SET-381 (Washington, DC: U.S.Government Printing Office, July 1988).

U.S.Congress, 1998 Transportation Hearings before a subcommittee of the Committee on Appropriations, House of Representatives. *105th Congress, 2nd session, Part 6.*

U.S. Congress, General Accounting Office, *Aviation Safety: FAA's New Inspection System Offers Promise, But Problems Need To Be Addressed,* Report No. RCED-99-183 July 7th, 1999.

U.S. Department of Defense. Standard Practice For System Safety. MIL-STD-882D, 10 February 2000.

U.S. Nuclear Regulatory Commission, *Fault Tree Handbook,* NUREG-0492, 1981.

Vos Savant, Marilyn. *The Power of Logical Thinking.* (New York: St. Martin's Press, 1996).

Watt, Kenneth E.F. *The Titanic Effect,* (New York: E.P. Dutton & Co., Inc., 1974).

Wells, Alexander T. *Commercial Aviation Safety,* (New York: McGraw-Hill, 1997).

Who Was Who In America, Vol.1, 1897-1942, (Chicago, IL: Marcus-Who's Who Inc., 1968), 7th Printing, pp.238.

Wiener, Earl. and Nagel, David. Eds., *Human Factors In Aviation,* (San Diego,CA: Academic Press, Inc., 1988).

Newspaper Articles, Press Releases, World Wide Web Information

Aviation Daily. "Team Effort Needed To Resolve Aviation's Problems." Sept. 22, 1999.

Aviation Week & Space Technology, "Air Travel In Crisis." October 25,1999.

ABC NEWS. "Construction Worker Zaps Miami Air Control Center." http://abcnews.go.com/wire/world/reuters19990709.

FAA News,"FAA Orders Immediate Inspection for High-Time Boeing 737's, Extends Inspection Order," FAA Office of Public Affairs, APA 57-98, May 10, 1998.

World Wide Web. Status of Boeing 737 Wiring Inspections As of June 16,1998," downloaded from the Web on 7/22/98: *www.faa.gov/apa/737iu.htm.*

Air Transport Association: *www.air-transport.org.*

FAA News, "FAA Announces Civil Penalty Settlement With America West," APA 90-98, July 14,1998.

FAA News, "FAA Acts To Increase Center Fuel Tank Safety," APA 92-98, July 23,1998.

FAA News, "FAA Continues Boeing 737 Wiring Inspections," APA 117-98, Sept. 28,1998.

FAA News, "FAA Issues Emergency Order on Boeing 747 Fuel Pumps," APA 143-98, December 3,1998.

FAA Safety Risk Assessment News. "Industry To FAA: Take A Systems Approach To Safety Assessments." Jan/Feb 1999.

Letter from the FAA Administrator, Linda Daschle, to Senator Ron Wyden, January 28, 1997.

Letter from Secretary of Transportation, Federico Pena to the Honorable Bud Shuster, Chairman, Committee on Transportation and Infrastructure, July 9,1996.

National Transportation Safety Board, "Transportation Fatalities Hold Steady in 1997; Highway Deaths Hit 42,000," News, SB 98-30, August 10, 1998.

Seattle Post-Intelligence Reporter. Www.seattle-pi.com/pi/local.

"Changes promised by the FAA haven't happened." February 23,1999. "Airline workers may face sanctions." January 26,1999.

"Safety last, FAA inspectors complain." March 5,1999. "Grounded by politics at the FAA. How safety inspector lost her dream job.

The Arizona Republic, "Record Safety Fine For Am West," July 15,1998.

The Los Angeles Times, Saturday, July 10,1999, Section: Part A. "GM Ordered to Pay $4.9Billion in Crash Verdict,"

The New York Daily News, 6/28/98 "Spilled Coffee Led To Near Miss;" *see also* "Blame for Jets' Near-Collision," *Newsday,* 6/29/98; "Coffee Spill Caused close Call at LaGuardia," *The Washington Post,* 6/28/98, P.A7; "Cause suggested for Near Miss," *The New York Times,* 6/29/98, PP.A17.

The New York Times, Editorial, "What's Wrong With Air Travel," June 19,1989.

The New York Times, July 5, 1998, p.A 12. See also: *The Washington Times,* July 4,1998, p.A4.

The New York Times, "Canada's Private Control Towers." October 23,1999.

The Times, August 28th 1999, p.8.

The Wall Street Journal, "Overbooked: A Crush of Air Traffic, Control-System Quirks Jam the Flight Lanes." September 1, 1999, p.1.

The Wall Street Journal, "Flying? No Point in Trying to Beat the Odds," Sept. 9,1998, p. A22.

The Wall Street Journal, "Sudden Acceleration Probe of Audis Finds No Defect," C8; 5, July 14, 1989.

The Wall Street Journal, Law, B6; 5, December 12, 1989. See also *The Wall Street Journal,* "Audi of America Agrees to Recall 5000-Model Cars," January 16, 1987.

The Washington Post, July 29, 1998, "Lois Gibbs's Grass-Roots Garden," p.D.1.

The Washington Post, July 9, 1998, "Nader Envisions a Tort Museum, Least Corvair and Company be Forgot," p.E.1.

The Washington Post, June 19,1999, "Carjack Victim Succeeds in Fight For Safety Latches in Trunks."

The Washington Post, July 10, 1999 "A $4.9 Billion Message," p.1.

The Washington Post, August 4, 1999, Metro, p.1, "Youth Killed by Dump Truck Will have a Legacy of Reform."

The Washington Post, October 30, 1999, p.1. "Boeing Delayed Fuel Tank Report."

U.S Department of Transportation, "Transportation Secretary Slater Unveils National Education Initiative," DOT 81-97, May 30, 1997.

U.S. Department of Transportation, "FAA Unveils Plan To Enhance Safety of Aging Aircraft Systems," DOT 180-98, October 1, 1998.

USA Today, "Pilots: Runway Crossings A Safety Hazard," November 13, 1998.

USA Today, "Report: Jet Safety Program Falls Short," July 7, 1999.

USA Today, "Proposal to cut hours truckers may work," August 6-8, 1999, p. 1.